VIETNAMESE INTELLIGENTSIA:
TYPICAL FIGURES

Hàm Châu

VIETNAMESE INTELLIGENTSIA: TYPICAL FIGURES

Thế Giới Publishers

First edition 2011
English translation
Printed in Việt Nam
VN-TG-3.66-1
ISBN: 978 - 604 - 77 - 0291 - 6

National Library of Vietnam Cataloguing in Publication Data

Ham Chau
Vietnamese intelligentsia: Typical figures / Ham Chau ; Transl.: Duong Kim Duyen... ; Quach Ngoc Anh revise. - H. : The gioi, 2011. - 386p. : phot. ; 21cm

1. Intelligentsia 2. Vietnam
305.509597 - dc14

TGF0042p-CIP

TABLE OF CONTENTS

FOREWORD

Vietnamese society's first contact with Western science dates back to the seventeenth century. In the form of objects such as clocks, telescopes, steamboats, firearms, and medicine, the technical applications of science were brought to the kings and lords of Việt Nam by Catholic missionaries and Portuguese, Spanish, Dutch, French, and English merchants, continuing through the French conquest (1862-1884).

During the first decades of the twentieth century, the concept of "science" was very vague, almost always synonymous with technology and technical invention. As such, Western science was considered as a product of crass materialism and "a tool of debasement of whatever rises up from the natural order," disregarded by Confucian scholars. Even Phan Thanh Giản, an intellectual mandarin sent on a mission to France, thought that France despite its civilization remained barbarous due to its lack of cultural foundations, namely Confucian doctrine.

With this mentality, science was unable to develop rapidly under the French colonial regime (1884-1945). Much different from British policy in India, for example,

French colonialism put stress on commerce rather than industrialization, giving priority in scientific education to the development of applied research rather than basic research. It was not until the 1930's that the Vietnamese intelligentsia became truly aware of the importance of science to the country's development.

This book, *Vietnamese Intelligentsia: Typical Figures*, intends to introduce typical Vietnamese scientists from different generations. It is divided into three chapters as follows:

The first chapter introduces the typical personalities from the first generation of modern Vietnamese scientists. Generally born in the families of Confucian scholars, they received university training in French, either in the former French Indochina or in France itself. They participated in the August Revolution (1945) led by Hồ Chí Minh from the very beginning and served in the two wars of resistance against the French (1946-1954) and the Americans (1965-1975). It is almost impossible to measure fully their talents and capabilities.

The second chapter presents the typical figures from the second generation of modern Vietnamese scientists. It was formed following the August Revolution and the two wars of resistance. They received secondary education in Việt Nam, continuing their higher education and obtaining post-graduate diplomas in the Soviet Union, China, and other countries in Eastern Europe. They are distinguished by their knowledge, and particularly by their expertise. They contributed to the resolution of practical and difficult problems during the American air-escalation in North Việt Nam and during the war in Việt Nam at the time (bombardments, blockade, armaments, communications and transport, logistics, etc.). Meanwhile, pure sciences were not neglected (mathematics, paleontology, parasitology, etc.).

The last chapter is dedicated to a presentation of several modern overseas Vietnamese scientists. Despite living far away from Việt Nam, they still have nurtured the burning ambition to contribute to their native land. These eminent figures have silently made considerable contributions to the development of Việt Nam in general and Vietnamese science in particular.

All the articles, presented in the ascending order of the scientists' birth-years, have been written by Hàm Châu, a journalist specializing in the scientific field. The composition of the texts is fairly diversified since we have gathered articles published at different times. They are not all the same length, the center of interest, or writing style. Still, we do hope that these articles will bring out particular faces among the personages of Vietnamese science.

Thế Giới Publishers

CHAPTER ONE

THE FIRST GENERATION OF MODERN VIETNAMESE SCIENTISTS

HỒ ĐẮC DI: SCIENCE AND CONSCIENCE

Professor Hồ Đắc Di was born on Buddha's birthday in 1900, the Year of the Rat. He was born into a noble family of the ancient capital city, Huế, that boasted of five daughters-in-law who were princesses and countless provincial chiefs and ministers. During the Vietnamese Revolution, President Hồ Chí Minh influenced this family by his example. Many members of this well-to-do family, such as Hồ Đắc Điềm, Hồ Đắc Di and Hồ Đắc Liên, along with their wives and children, were inspired to voluntarily leave their convenient city lives behind to join the revolution, sharing half a bowl of rice and a thin blanket with soldiers and compatriots in the middle of the forest. It was President Hồ Chí Minh's devoted, pure, and noble life that converted doctor Hồ Đắc Di – resident physician of Paris hospitals, researcher who announced thirty-seven valuable scientific works starting at a rather young age, and the first and only Vietnamese university professor prior to the August Revolution – to the cause of North Việt Nam. He wrote, "Yesterday, I was consumed by my profession and had a passion for technology, but tomorrow I will eat vegetarian food and lie on the ground with the masses."

It was this famous intellectual, son of a duke who was a member of the Đông Các Academy and a physician educated in France that sat under the eaves of his residence in Vân Đình on the night of December 19, 1946, looked towards Hà Nội, saw the light from the blazing fires, listened to the perpetually roaring gunfire, and uttered, "So the revolution has broken out!" On that very night, moved and inspired by the courageous sacrifice of the suicide combatants protecting Hà Nội, the professor hurriedly wrote the following words in a notebook by the light of an oil lamp whose flame flickered in the cold wind, "If only there were a life in which every second that passed were as meaningful as a thousand. ...If there were such a life, I would be willing to devote all my life to it, to accept loss and sacrifice even if it was only for the sake of my descendants. Anyone who has lived a life of oppression and disgrace, experienced long nights of slavery or even suffered miserable and conscience-stricken days would certainly rush into the whirlwind of revolution once its bright light had illuminated their soul..."

Even years later, he could remember clearly the bright firelight of the earth-shaking whirlwind of revolution – the flame of a brave Hà Nội which stood up and burnt itself to become immortal. This firelight of resentment and hatred burned on the night of the nineteenth – the first night of the national salvation struggle, whose participants "would rather sacrifice everything than lose the country or become slaves."

It is the above-mentioned diary entry written in subtle French on the night when "Hà Nội was blazing and full of smoke" that Professor Hồ Đắc Di later used to start the memoir "Let the Soul Be Easy" (1974). He never forgot the strong emotions he felt that night.

In June 1998, the Việt Nam Culture and Information Publishing House published the first volume of the long book titled *Lịch sử và văn hóa Việt Nam - Những gương mặt trí thức* (History and Culture of Việt Nam - Faces of Intellectuals). This book details the lives of noble intellectuals from the feudal times, great intellectuals, and personalities from the modern era and patriotic intellectuals from the Hồ Chí Minh age - including Hồ Đắc Di. Hồ Đắc Di was not only well-known in the medical field, but was also a famous man of the nation.

Hometown and Family

Professor Hồ Đắc Di's privileged upbringing gave him more access than most people to the best intellectual and cultural education available. His paternal grandfather was Marquis Hồ Đắc Tuấn, a Confucian scholar who married Princess Thức Huấn, daughter of Tùng Thiện Vương. Hồ Đắc Di's father was Hồ Đắc Trung, Duke of Khánh Mỹ. He was also a Confucian scholar and first married Princess Á Nam and then married Châu Thị Ngọc Lương. Because he was the son of a duke, during his nearly thirteen years of studying abroad, young Hồ Đắc Di was affectionately dubbed *Prince Hồ Đắc* by his French friends.

Hồ Đắc Trung (his father) worked as Minister of Education, as a high - ranking mandarin of the Đông Các Institute, and as a member of the Privy Council. As Minister of Education, he helped start Dục Thanh School in Phan Thiết Town, which was the most progressive private school in Central Việt Nam at that time. It was at this school that young Hồ Chí Minh would teach elementary school students under the name Nguyễn Tất Thành before he left the country.

All of Hồ Đắc Trung's children achieved high positions in society. His eldest son, Hồ Đắc Khải, obtained a bachelor's degree in Confucianism, married Tôn Nữ Thị Khâm (elder sister of Professor Tôn Thất Tùng, who was eventually a close colleague of Hồ Đắc Di), and worked as Minister of Finance, Senior Academician, and member of the Privy Council. After the August 1945 Revolution, he lived in Sài Gòn and worked for neither the French nor the Americans.

Unlike Hồ Đắc Khải, who studied Sinology, his brothers received Western educations and obtained high degrees rare among Vietnamese at the time. Hồ Đắc Điềm earned his Ph.D. in Law; Hồ Đắc Di became a medical doctor; Hồ Đắc Liêm became an engineer of mineralogy; and Hồ Đắc Ân graduated with a Ph.D. in Pharmacy.

Hồ Đắc Trung's daughters also obtained high positions in society: Hồ Thị Huyên married Ưng Úy and gave birth to scholar Bửu Hội, who worked at National Center of French Scientific Research; Hồ Thị Chỉ was the most beloved concubine of Emperor Khải Định of the Nguyễn dynasty; and Hồ Thị Hạnh (also known as Bonze Diệu Không) went on hunger strike at Temple Ấn Quang to oppose the Sài Gòn administration's oppression of the Buddhist circle in 1963. Hồ Thị Hạnh was also a poet who translated complex books such as Buddhist scriptures from Chinese characters into Vietnamese.

Hồ Đắc Di – the Student

After they completed their years of primary school in Huế, Hồ Đắc Di and his brother, Hồ Đắc Ân, were sent to Hà Nội, where they attended the Paul Bert Secondary School (now known as Trưng Vương Secondary School, located on Hàng Bài Street). As all of his subjects were taught in French, and he had few opportunities to speak in

Vietnamese, Hồ Đắc Di became accustomed to expressing his thoughts and feelings in French, a habit that remained with him in his later life, when he did most of his writing in French, including the articles he wrote during his revolutionary days by Quẵng Stream, Chiêm Hóa District, Tuyên Quang Province. Many of his articles that were printed in Vietnamese books or newspapers were only translations which could not do justice to his refined and subtle talent or his interesting way of playing with words in the original language.

After World War I ended in 1918, Hồ Đắc Di set sail for France to continue his studies there. He finished high school in Bordeaux in 1920, after which he took the entrance exam for the prestigious Medicine Faculty of Paris University. It was very difficult for foreign students to achieve the level Hồ Đắc Di reached, but his proficiency in French helped him to pass even the resident-physician exams, a feat which only 80 out of 1,200 candidates achieved. Hồ Đắc Di proceeded step-by-step until he was a resident physician studying master surgery techniques in the Paris hospitals.

Hồ Đắc Di managed to receive the highest possible grades in the school of French medicine – one of the most progressive in the world at the time. He was taught clinical and surgical skills by famous teachers at Cochin Hospital, Saint Antoine Hospital, and finally, Tenon Hopital. Professor Fernand Widal taught him methods of quick and exact diagnosis through clinical observation. Professor Gernet and Professor Moulonguet taught him surgery. Professor Gernet, a person with a wide range of knowledge, a passion for books, and an artistic soul, had a profound influence on Hồ Đắc Di, his assistant, in other ways as well. It was through him that Hồ Đắc Di gained a reputation in the French intellectual circle. Hồ Đắc Di, a Vietnamese resident physician in Paris hospitals, and Fujita Prince, a Japanese artist, were two young talents paid particular attention by Paris' elite at that time.

Hồ Đắc Di became particular friends with Ève Curie, daughter of Nobel Prize winners Pierre and Marie Curie, who often invited him to visit their house on Sundays. Ève was sentimental, good at prose and verse, and fond of music and make-up. She liked to listen to Prince Hồ Đắc play the violin and appreciated his graceful, humorous personality, and his knowledge of philosophy, literature, and the Western arts. Many years later, Hồ Đắc Di still kept a picture of Ève listening to him play the violin at Marie Curie's house.

His wide knowledge of philosophy, literature, and Western arts enabled Hồ Đắc Di to develop profound philosophic thoughts and highly polished writing skills that would appear later in the articles he wrote in the Việt Bắc forest. His passion for music remained obvious and he often mentioned Beethoven and Tchaikovsky.

In the 1920s, many Vietnamese students went to study in France, many of whom were intelligent and studious. However, most of these students were used to passive learning through books by means of careful, detailed study, and memorization, rather than by any real creativity or investigation. The young student Hồ Đắc Di was one of the first Vietnamese students to become fully aware of the significance of scientific research in universities and realized that "teaching and research are twin brothers and that the magisterial chair is only the anteroom of the laboratory" and that "universities were not only the place for teaching formed sciences but also the place for forming sciences."[1]

Thanks to his passion for study, in his doctoral thesis Hồ Đắc Di proposed a new method of gastric surgery which was named after him, acknowledged, and quoted in monographs and trade magazines. It was a method of intercommunication between the gastric system and the duodenum, aimed at curing pylorus-stenosis without removing the stomach, as was done previously. Hồ Đắc Di's new method was adopted in many countries.

On December 13, 1941, in a speech at the French Medicine Exhibition in Hà Nội, Professor Hồ Đắc Di recalled his years in Paris, "I am moved whenever I think of my youthful memories of the time when I was hard-working, studious, and happy. I experienced the passionate Paris lifestyle and tasted joy and sweetness of both heart and mind."

However, during the years he studied and socialized carelessly, important events were happening in his homeland. Vietnamese Emperor Duy Tân was exiled to Reunion Island for his French resistance, Phan Châu Trinh went to France and met with workers and students, while Phan Văn Trường wrote his Ph.D. law thesis in Paris and publicly praised the Russian October Revolution. In particular, it was during this time that Nguyễn Ái Quốc (later known as Hồ Chí Minh) proposed his eight claims at the Versailles Peace Conference. The statement was written at Rue des Princes with the title of "Le Droit des nations" (Rights of Nations) and signed "Nguyễn the Patriot."

1. Extract of "Mission and Role of the University," November 1949.

These events did not go unnoticed by Hồ Đắc Di. While he was still in high school in Bordeaux, the 19-year-old Hồ Đắc Di received a leaflet containing "Le Droit des nations" from a Vietnamese worker. Despite his ignorance of the complicated political issues behind this document, he admired the extraordinary courage of its writer. When he moved to Paris, he joined the Vietnamese student club, which met at 15 Sommerard Street, Latin block. Hồ Đắc Di actually had two encounters with Nguyễn Ái Quốc when the revolutionary initiated meetings with Vietnamese students. Though he admired this slim, bright-eyed man, only ten years his elder, Hồ Đắc Di did not then imagine the profound influence this man would have on his own life and on his entire nation.

Hồ Đắc Di's Professional Years

In 1931, when he was working as an assistant to Professor Gernet at Tenon Hospital, Dr. Hồ Đắc Di was called back to his hometown to get married. He was excited to use what he had learned during the thirteen years he had spent in France to help his people. However, he soon became frustrated by the injustices he experienced working at Huế Hospital, where French doctors discriminated between patients and treated Hồ Đắc Di and his Vietnamese colleagues as inferiors. Hồ Đắc Di was tempted to return to France, but was persuaded not to do so by his father, who wanted him to stay to prepare for his upcoming wedding.

He went to Hà Nội and sought out Leroy des Barres, then Director of Hà Nội University of Medicine and Pharmacy, who had formerly worked as a resident physician in the French hospitals. Barres invited Hồ Đắc Di to teach at the university, but Di insisted that he would not take the job unless he could also work as a surgeon. As there were only two licensed surgeons in Indochina at the time, Hồ Đắc Di was given a dual position as lecturer at Hà Nội University of Medicine and resident surgeon at Phủ Doãn Hospital (then known as the Hospital of the Protectorate). He often surprised Vietnamese students by working quickly without consulting his French superiors – something unheard of at the time.

Once in an emergency case, he asked a male nurse to prepare tools immediately. The nurse was so surprised that he stood still for a while and then stammered, "Did you ask the high boss?"

To which Dr. Hồ Đắc Di replied, "There is no need to ask. I will do it."

His student was shocked; he couldn't believe that any "Annam" (Vietnamese) person could be better than "the high boss," Leroy des Barres. At that time, Hồ Đắc Di was only thirty-two years old, but he was called "old" by the male nurse and later by many other people. Whenever he mentioned it, Professor Hồ Đắc Di joked, "I was old so fast. I became 'old' at the age of thirty. It is so funny; I am the old but young guy."

Hồ Đắc Di worked at Phủ Doãn Hospital from 1932-1938. During this time, he became well known for taking charge, making quick, accurate diagnoses, and teaching his students in a concise, easy-to-understand manner. He often taught them as follows, "The medicinal art includes a subtle spirit called clinical sense. [...] It is this clinical art which leads us to distinguish the subordinate from the essential among the morbid symptoms, to judge not from a particular but from the whole. It points out to us what is, in a given case, the therapy adequate to the morbid necessities, adapted to the biological differences and the social conditions of the individual. [...] The close relationship of medicine and the social sciences makes him (the physician) think it his duty to make himself a master of a wide general culture, which is to make of himself something different from a skilful artisan educated at a vocational school. This formation asks for the development of all his faculties, intellectual and moral as well as scientific and artistic. [...] The artist is above all a sensitive man, and so is the real physician whose value, just like that of the painter or the musician, is proportional to his sensibility. It is by uniting science and culture that the physician proves to be worthy of his noble profession."[1]

When World War II broke out, Dr Meyer-May, a former Jewish French resident physician of Paris hospitals, came to Hà Nội from Delhi via Bangkok to see Hồ Đắc Di. On Dr. Hồ Đắc's advice, he moved to work at Hà Nội University of Medicine as he was one of Professor Antonin Gosset's students and one of the world's best surgeons. Famous for good surgical qualifications and good-looking surgery scars, he later cooperated closely with doctors Hồ Đắc Di and Tôn Thất Tùng in surgery as well as in scientific research. No one expected that he would later secretly join the De Gaulle group against fascism.

In 1938, Hồ Đắc Di's student, Tôn Thất Tùng, was the only student to pass the very difficult examination and become a resident physician in the hospitals of Hà Nội. Under Hồ Đắc

1. A speech made by Prof. Hồ Đắc Di on the opening day of the Vietnamese National Medical Conference, 1949.

Di's tutelage at the Phủ Doãn Hospital, Tùng soon became an adept surgeon, and Hồ Đắc Di delegated much of his nighttime emergency cases to the younger doctor. Later, when he became a world-famous scholar, member of the Paris Royal Academy of Surgery, and was awarded the Lannelongue Prize – an international prize in surgery – Professor Tôn Thất Tùng dedicated every one of his scientific publications to Professor Hồ Đắc Di as follows: "Dedicated to Old Di, my first teacher." In conversations with young scientists, Professor Tôn Thất Tùng said, "To do science is to reconsider issues. If not, there will be nothing to be sought and nothing to be invented. Old Hồ Đắc Di, my teacher, often taught me, 'Science is the revolt of thoughts.' I remember and follow those teachings." The two men worked together for the rest of their lives.

During the time before the August Revolution, the Indochinese colonial administration prevented Vietnamese from obtaining high positions within the hospital or university medical systems. Thus, at Phủ Doãn Hospital, the business was controlled by the French directors, although the three Vietnamese doctors, Hồ Đắc Di, Vũ Đình Tụng and Tôn Thất Tùng did most of the professional work. This discrimination contributed to the smoldering dissatisfaction of the Vietnamese physicians and students. Hồ Đắc Di later expressed this frustration when he wrote, "Something began to worry me. Though my life appeared normal, I suffered inwardly. [...] Misfortune opened my eyes and what I had only had a glimpse of became a tangible reality. I was only the citizen of a conquered country."[1]

When the Japanese entered Indochina, the French became subject to the Japanese fascists. The Allied airplanes

1. Memoirs "La voie que j'ai choisie."

bombed Hà Nội, causing casualties to many civilians who were sent to Phủ Doãn Hospital. Doctors Hồ Đắc Di, Vũ Đình Tụng, and Tôn Thất Tùng, together with resident students and nurses, worked day and night on seven operating tables.

In addition to his medical work, Hồ Đắc Di wrote and researched prolifically during the time he worked at Phủ Doãn. He was the first person to research trauma shock in Indochina and he published works on this subject in *Revue medicale française d'Extreme-Orient* (French Medicine Magazine in the Far East). He was a popular and prestigious writer whose many articles on peritonitis surgery and gastric-duodenum ulcer surgery in northern Việt Nam were also published in this magazine. In addition, he received publication offers from medical journals such as *Academie de Chirugie de Paris* (Paris Royal Academy of Surgery) and *Gazette medicale de la France d'Outre-Mer* (Overseas French Medical Paper). His research findings in acute edematous pancreatitis and technical treatment on the side-effects of peritonitis due to typhoid and gallbladder perforation opened new doors for his colleagues and students. Moreover, he initiated a caesarian technique that was published in a French medical journal in Paris. Throughout his works, he focused particularly on regulating the characteristic diseases of his impoverished, underdeveloped tropical country.

Many of Hồ Đắc Di's works during the time he worked at Hà Nội University of Medicine and Phủ Doãn Hospital were written in cooperation with prestigious foreign professors like Huard and Meyer-May as well as with his close Vietnamese partners like Tôn Thất Tùng and Vũ Đình Tụng. Records from Hà Nội University of Medicine show that only twenty-one of Dr. Hồ Đắc Di's known works have been found.

It is not enough to say that Hồ Đắc Di was just a medical thinker who effectively trained Vietnamese medical personnel during his lifetime; indeed, he was so much more. He was the first Vietnamese surgeon and an international-level researcher. Thus Dr. Hồ Đắc Di was a rare man among Vietnamese of his time. Because of his talent - as well as to relieve the dissatisfaction of Vietnamese physicians and students - not long before the August 1945 Revolution, the Professors' Council of Hà Nội University of Medicine decided to elect Dr. Hồ Đắc Di an associate professor, eventually promoting him to full professor. For similar reasons, they appointed his student, Dr. Tôn Thất Tùng, to the position of lecturer.

Difficult but Inspirational Days

How could people like Hồ Đắc Di and Tôn Thất Tùng who grew up in noble families and had high positions in society ardently join a revolution? How could people of such high caliber leave Hà Nội's streets and their convenient villas with full facilities and, together with their wives and children, follow evacuees flocking to villages with their entire fortunes wrapped up on rickshaws in preparation for the national resistance day? How could they have the spirit and perseverance to maintain the medical school and train four times more doctors than there were before the war during eight years of living in the middle of forests or in small mountain villages in Chiêm Hóa, Trung Giáp or Lang Quán?

The only possible explanation is that these men possessed great consciences and respected human dignity. Old people say, "a person of high character, or one who thinks highly of benevolence, would not veer from benevolence to seek life and would rather kill oneself to keep one's benevolence." On a wad of rough, stained, small, bamboo papers, hand-made in Việt Bắc during the anti-French age, I still can read these words that Professor Hồ Đắc Di wrote in purple ink, "All preferred to die a glorious death than to live a humiliated life. All died for widows' tears and orphans' gloomy childhoods, for refugees' grievous groups and burnt down villages, for the wasted fields and destroyed crops, for the destruction of millions of beings and of the fruit of an immense creative labor."[1] Thus, it is clear that the legendary individuals who pioneered the way for future generations were inspired and strengthened by hope and their strong characters.

After the August Revolution, President Hồ Chí Minh and the government assigned Professor Hồ Đắc Di many important responsibilities. They made him Director of the University of Medicine and Pharmacy, General Director of Higher Education Department, General Inspector of Medicine, and Director of Đồn Thủy Hospital. Professor Hồ Đắc Di recalled, "I devoted myself, body and soul, to work, and my cares vanished without my noticing it. The dreams I had been nurturing for so many years were going to be realized." After having experienced a few short days of freedom, Dr. Hồ Đắc Di found French rule even more intolerable when the tanks and armored vehicles of the French expeditionary troops rolled back into Hà Nội.

Today, it is difficult to imagine how the Hà Nội University of Medicine and Pharmacy could keep teaching students during the hard days of anti-French resistance,

1. Speech made by Prof. Hồ Đắc Di on the opening day of the Vietnamese National Medical Conference, 1949.

particularly during 1947-1950 when the Chinese revolution failed and provinces along the Việt Nam-China border were not liberated. The time of the French resistance war (more widely known as the First Indochina War) was a difficult time for the various universities. Many tried and failed to remain open and teaching. Others had to move.

During the first days of the resistance, Professor Nguyễn Xiển opened a class of general mathematics including about ten students who initially studied by sending lessons, back-and-forth, then gathered and studied in Đại Điền Village, Tam Dương District of Vĩnh Yên Province. In October 1947, the French attacked Việt Bắc, forcing those students to escape to rock caves. The class was postponed and they returned to their respective hometowns.

In a free zone in Nam Đàn District, Nghệ An Province, where French troops did not enter, Professor Nguyễn Thúc Hào, former acting Director and General Secretary of Hà Nội University of Science, also opened a program of general mathematics which lasted longer - from 1947 to 1951. However, this program only taught at the first year level; it was not a complete program for a science bachelor's degree. From 1951 to 1954, a junior college, and then a senior teachers training school were opened in the rear areas of Thanh Hóa Province where there was little violence. Professor Nguyễn Thúc Hào was appointed member of the director's board at these two schools by Ministry of National Education with Trần Văn Giàu, Đặng Thai Mai, Đặng Xuân Thiều, and Nguyễn Mạnh Tường.

Not until 1951-1952 were Basic Science School and Senior Teachers Training School opened at Central Dormitory Area in Nanning (China), where the French could not attack.

Famous professors like Lê Văn Thiêm, Nguyễn Xiển, and Ngụy Như Kon Tum also worked there.

This period was no less difficult for the Hà Nội University of Medicine and Pharmacy, which was now headed by Professors Hồ Đắc Di and Tôn Thất Tùng. When the school was still based out of Hà Nội, Professors Hồ Đắc Di and Tôn Thất Tùng – along with their respective families – were forced to move around frequently in the provinces surrounding the city. The two professors moved from Vân Đình to Sơn Tây, then to Phú Thọ, and up to Tuyên Quang, accompanied by their families and about ten voluntary medical students. At each new location, they had to set up operating rooms, arrange and sterilize surgery tools, and evacuate casualties. Professor Hồ Đắc Di was in charge of doing surgery cases throughout Kim Lũ, Cự Đà, Vân Đình, Hòa Xá, and Đốc Tín, and receiving wounded soldiers from Hà Nội's battlefields.

Eventually, when the French took over Hà Nội, the university hospital was forced to withdraw to the Việt Bắc revolutionary military base, though not without some regrets. Professor Hồ Đắc Di remembered the trip, writing, "We arrived at Quẵng Stream one evening at the end of May 1947. The region was mountainous and deserted. Nothing but green foliage containing mosquito nests and wild beasts' dens could be seen. There was a heavy smell of rotten leaves. The leeches raising their suckers everywhere certainly did not allay my apprehension.

Before me, the Quẵng Stream, swollen by flooding, carried muddy water that struck the stony banks with a heavy rumbling nose. There was one lonely house: the ferryman's. I looked at my wife and children, then at the jungle around, and my heart filled with a dim sadness. The voice of my conscience had induced me to take the way of resistance. But, strictly speaking, it had not wiped out all anguish from the bottom of my heart. I and my companions immediately started to work. And the things that had been taking place were much better than I had thought."[1]

The professors and students set up the university hospital on a hill in Ải Village, Tuyên Quang Province. Ải village was about five kilometers away from Chiêm Hóa District of Tuyên Quang Province. Local people often traveled by raft along this canal. On the two banks were scattered small villages of Tày and Dao ethnic minorities and the surroundings were nothing but hills and vast forests. With available bamboo, the doctors, together with their students, learned from local people how to make operating rooms, treatment rooms, and laboratories. Since it was located on a hill, the hospital was called "hospital hill" by local people, and the place retained the same name even long after it had closed.

After they had stabilized the hospital, accommodations, and study area, the school organized graduation exams for the 1946-1947 school year. Eight out of fifteen sixth-year students passed the clinical exams in surgery and obstetrics, graduated, and were sent back to their previous positions on the battlefield.

On October 6, 1947, the university held a ceremony in Ải Village to initiate the new academic year. In addition to the university's professors, students, and staff, and their families, many important government officials and local villagers attended this momentous event.

1. Quoted from Hồ Đắc Di memoirs "Let the Soul Be Easy," Y học (Medicine) Publishing House, Hà Nội, 1999.

After pondering for a long time about medicine and the mission of a medical university, thoughts that combined with passionate feelings about those first days of the resistance war, Professor Hồ Đắc Di sat down in the quiet of the Việt Bắc forest to write a speech appropriate for the ceremony. The resultant speech, which he composed in French, can be seen as a patriotic declaration of the mission of university education during the Vietnamese revolution. The speech has now been translated into Vietnamese; however, the language is not well-polished and thus does not do it justice. However, Professor Hồ Đắc Di's words were greeted with enthusiastic applause from his French-proficient audience in the Việt Bắc forest.

In his speech, Professor Hồ Đắc Di, now simultaneously Director of the University of Medicine and Pharmacy and General Director of Higher Education Department stated, "Sensitive to the rhythm of the fatherland, to its pains as well as to its joys, our spirit strained toward independence, our eyes looking fixedly at a victory in which neither the longing for conquest nor the desire for domination will take part; rather, inspired by the hatred of servitude and the love of freedom, the university will serve by duty and without pride but with dignity because victory must be gained and the university wants to help achieve it. To all those who fight for the fatherland's freedom, the university is proud to send its fraternal salutation and takes a solemn vow to see to the integrity of their intellectual patrimony."

What was surprising at the time was when Professor Hồ Đắc Di stated the following simple principle of university education: to know the truth and love the beautiful in order to do good. What is to know the truth? It means "rising above particular facts, showing the importance of general laws, learning to reason with an infallible dialectic logic and showing oneself absolutely loyal and impartial before facts."[1] In other words, being true makes one beautiful. Once we know the truth and love the beauty, we should act in a good direction, or in other words, do good deeds. A scientist should have not only a scholar's brain but also a kind heart.

Not many people in the year 1947 had a vision extensive enough to realize that "the university is not only an amalgam of schools, a center of distribution and repartition of knowledge and techniques, but also a center for research. [...] In the scientific order, progress postulates and demands this intimate and indispensable collaboration without which one can only build on sand."[2]

Professor Hồ Đắc Di believed a university professor's mission was "to distinguish young students who like scientific research, to help them to develop a spirit of observation, criticism, curiosity, and intuitive imagination while still recognizing the need for knowledge. [...] The master is all the worthier the further he is left behind by his own pupils."[3]

Professor Hồ Đắc Di focused on the spirit of the university and the specific characteristics that distinguish between university and elementary school, high school or vocational school. According to him, the main distinction was the warm and democratic relationship between teachers and students based on trust and friendship. Only in such a comfortable atmosphere could brains of scientific criticism – the most beautiful bunch of flowers in man's knowledge – bloom freely.

1. *Mission and Role of the University.*
2. *Mission and Role of the University.*
3. *Mission and Role of the University.*

Professor Hồ Đắc Di put his hope in the young generation, "Scientific research is often carried out by young people. So in any era, most inventions come from the youth..."

Not long after that, the professor wrote, "The difference between creation and practice is that we can see one but not the other. Positive and pure continuation one side's tradition is contrary to the selective and critical activities of the other side."

It seemed so hopeless to make students enhance their capability of observation and do scientific research in the middle of Việt Bắc forest but upon thinking it over, we will see that is was strangely wise. Thanks to this approach, many of Hồ Đắc Di's students later became creative scientists, including Nguyễn Huy Phan, Phạm Khuê, Đặng Hồng Vân, and Hà Văn Mạo. The speech for the opening ceremony concluded with the following brave words, "On the dial of history, the clock has struck the hour of liberation. It sounds the assembly of all the energies to carry out this supreme endeavor which is to bring freedom and independence back to us, to release us forever from the chains of servitude of this sacred land which was yesterday the cradle of our childhood and will be tomorrow be the refuge of our eternal rest. [...] Where can we, gentlemen, find ground for hope, if not in the comforting spectacle shown to us by the life and work of these young people ready to sacrifice?[...] and where can we find reason for confidence, if not in the inspiring observation which is brought to us by the life of this man, made of total abnegation and sacrifice for almost half a century, the man to whom an extraordinary destiny has befallen and who is conducting our orchestra?"

It's hard to believe that the words Professor Hồ Đắc Di spoke modestly from a bamboo house decorated with local flowers beside the desolated and solitary Quẳng Stream can still move us to tears today.

Moments of Choice

No sooner had the new school year started than the French launched an attack on Việt Bắc. President Hồ Chí Minh wrote a letter warning the professors and students, who hid deep in the forest. He wrote, "The resistance war has now taken a new and tense step as we predicted, but the sooner it comes, the better it is." He also said, "Do not fear the enemy but absolutely do not underestimate them;" in the face of challenges, show your spirit, "not turning back despite hundreds of winding roads" and "patiently enduring hardship and misery." He instructed them to make their plans carefully and "keep all the families and students safe." Finally, President Hồ wrote, "please give my wishes to your wives, my kisses to your children and my regards to the students" in his post script.

The French were stationed in Chiêm Hóa District for one month. They searched the surroundings every day and aggressively advanced in the direction of Ải Village where the school was located. Eventually, the French set fire to the practice hospital of the University of Medicine, killing one patient. This scenario became all too common as the French continued to invade and destroy parts of the Việt Bắc forest area. The professors and students opted not to leave, however, hoping that the French would soon leave and allow them to continue their work. In the meantime, they were forced to go into hiding during the days, only returning to the village at night.

Finally, the French parachute troops were blockaded and beaten badly. The North Việt Nam Army set their canoes and war ships on fire, and decimated 27 of their trucks and armored vehicles. By late December, 1947, French troops were forced to withdraw back to Cầu Đuống. After the Autumn-Winter operation failed, the French troops did not dare to risk another attack on the Việt Bắc resistance zone. Everyone – including the ethnic minorities and the professors and students of the medical university – returned to their villages in time to celebrate the Lunar New Year's festival in 1948.

However, prior to their withdrawal, French troops had set fire to the school's facilities, destroying many buildings, books, and medicines. Professor Hồ Đắc Di met with President Hồ Chí Minh to discuss what should be done about the school. The President told the professor that the school must remain open and promised that the government would support its operation.

Professors Hồ Đắc Di and Tôn Thất Tùng decided to move the school back to the midland region near the battlefield in Trung Giáp so as to make it easier for the army to be served and the students to practice. Soon after, they succeeded in setting up a new school and practice hospital made of fresh bamboo trees in Trung Giáp. Soldiers – including those from areas temporarily occupied by the enemy – and local people once again came to the school for medical treatment. Stores and inns sprang up around the hospital. The school set up an advance station in Phú Hộ – two kilometers from Trung Giáp – to receive wounded soldiers and give them first aid treatment. However, one morning the French had their parachute troops land there, causing physicians and students to flee hurriedly into the forest. The entirety of the school's property was once again set aflame by the French.

The two professors decided to return to Chiêm Hóa and build a long-lasting base in Ải Village, acknowledging that neither a university nor a hospital could work effectively if it had to be constantly on the move. Once again, the teachers and students pooled their efforts to set up new buildings in the middle of the forest. They rewrote textbooks and re-opened the practice hospital. Võ Như Tỷ, a student who had previously studied architecture, worked with Professor Tôn Thất Tùng to design beautifully-shaped and well-lit patient and operating rooms suitable for a hospital. Lacking suture threads, they replaced them with threads taken from the ropes of parachutes that had been taken from the French. When he needed light to operate by, Professor Tôn Thất Tùng had a student pedal a Peugeot bicycle to power the generator. The hospital also managed to procure a diesel engine and an X-ray projector. Its members successfully produced penicillin derivatives.

During the eight years of the Indochina War, the University of Medicine had to move thirteen times, setting up new facilities each time. The students who studied in Việt Bắc later became academicians, professors, ministers and institute directors in the medical field all over the country.

Reflections in Việt Bắc Forest

While in the Việt Bắc forest, especially after the French attack on Chiêm Hóa that destroyed all the available books, Professor Hồ Đắc Di wrote prolifically. Living a quiet, simple life in the forest gave him time to think and write on many subjects, free from the hustle and bustle of the base and material demands. Thus, the drafts that have been found do not consist only of fiery, patriotic words in favor of the resistance against the French, but also lucid ideas about university education, scientific research, the medical field, the mission of physicians, and his own philosophical thoughts.

Professor Hồ Đắc Di once said, "The psychological activities of a man can be compared to a resonant engine which only oscillates when stimulated by the same-frequency oscillation of another object. When this happens, both of them will oscillate at a much bigger amplitude."

When the August Revolution took place, many Vietnamese in the nation joined the resistance. In November, 1948, after the victory of the 1947 Autumn-Winter Việt Bắc Operation, Professor Hồ Đắc Di wrote the following words by Quẳng Stream, "Unprecedented events follow one another with extraordinary speed. They are of such a nature, of such breadth and endowed with such character that it becomes impossible for anyone that really loves the great, the noble, the beautiful to remain insensible. The fatherland's luster, unknown until now, rises to such a height that our people transcend themselves and our nation is filled with amazing deeds. Now we are proud of being Vietnamese, while even a short time ago we were almost ashamed of it because we were stateless persons." (*Mission and Role of the University*)

We don't hesitate to quote the professor's words as they were very beautiful and emotive phrases from the bottom of his heart which are still necessary for future generations.

"Thus we were faithful to this past action as if it is by going to the sea that the river is faithful to its source. [...] It is by becoming imbued with these virtues that the university has been able to see in the present the accumulated force of the greatness of the past and the token of hope of the future. An opportunity turned up for us to force destiny, to break the chains of servitude, and we rose to the occasion. Our resistance, which is measured by our hatred of servitude, has revealed itself equal to the events and is made on the scale of man and of the nation. Resolute in our resistance, we will escape not by 'the force of the bayonet' but by the force of freedom." (*Mission and Role of the University*)

Since the very early days of resistance war, Professor Hồ Đắc Di distinguished clearly between French people and its culture and the invaders and imperialists. He admitted that the 1789 French Revolution had made many Vietnamese people fully understand the value of freedom. Despite that, he questioned, "Does this country – which was the first to inscribe the words 'Freedom, Equality, Fraternity', in gold letters on the frontispiece of its institutions – continue to practice a policy of domination, to assert itself as a master and to sow hatred? What! Has the Cartesian logic of which this country is so proud deserted its own fatherland, which is also a victim of the Fascist persecution?" (*Mission and Role of the University*)

Professor Hồ Đắc Di affirmed that Vietnamese people wished to enter the house of independence formally, by "the front door, not by a leak in the side door" and that they were looking forward to the rights of freedom which only real independence was able to satisfy.

However, Professor Hồ Đắc Di's theses written in the middle of Việt Bắc forest were not only the fiery words of a patriotic intellectual. Those of the old cockroach-gnawed drafts written by Quẵng Stream which are still preserved today also contain lucid, pioneering ideas about university education, scientific research, the medical field, and the mission of physicians as well as his own philosophical thoughts. He expressed all of these surprisingly profound insights succinctly, in stately, poetic language.

After the French attack in Chiêm Hóa, the University of Medicine lost many books and textbooks. The remains were once again set on fire when the enemy's parachute troops surrounded and destroyed the school set up in Trung Giáp. Without available books for studying and quoting, Professor Di could do nothing except write down his thoughts. He lived a frugal, quiet life in the middle of the forest by the stream in harmony with flowers and trees, just as scholars did in the old times, distancing himself from scrambling and hunting at jet speed for the illusion of happiness based on material demands. Maybe it was this simple lifestyle that enabled him to sublimate his philosophy.

Since an early age, Professor Hồ Đắc Di had a dialectical perspective on many matters. He saw the unity of two opposing sides and analyzed both sides, rather than judging one inflexibly. Over half a century ago, Professor Hồ Đắc Di mentioned the relationships between inheritance and development, between studying what was known and inventing what was unknown, between conscience and intelligence, between the arts and the sciences, between culture and technology, between generalization and analysis, between intuition and rationalization, between theory and experiment, between induction and deduction, between indefinite and definite, between relative and absolute, and between the whole and the part. It would be difficult to understand Professor Hồ Đắc Di fully in the limited sphere of his professional knowledge, because Professor Hồ Đắc Di was not only a medical professional but was also a sensitive and erudite person, a humanistic thinker, and a modern scholar who "was not seduced by wealth, shaken by poverty or persuaded by military power."

In 1949, Professor Hồ Đắc Di wrote the following words about medicine at the National Medical Conference in Việt Bắc, "Those who want to link their fate to this difficult job must determine that, in this important mission, the heart is as important as the brain." Similarly, in the years following 1949, Professor Hồ Đắc Di warned about the danger of "bureaucratizing" physicians, a practice that has become rather popular over the past half century, "He ends by becoming a civil servant, asphyxiated by red tape and administrative rules – a bureaucrat. When this is the case, his presence becomes a superfluity and his absence a lack; he is merely compelled to be present for a fixed number of hours. If he does not know how to react, he will end up forgetting even the names of illnesses that he learned in his previous life as a medical student. People have succeeded in functionalizing the physician but not illness. A perforated ulcer, a pernicious attack, a dystocial delivery – all of these still ignore the opening-and closing-times of offices. Very often, all these little miseries – little for the physician, but great for the patient – prefer night to any other time. They appear at two or three o'clock in the morning, an hour when the bureaucrat is assured of finding perfect security between the arms of Morpheus or of Venus. How retrograde illnesses are! When will they understand that one must serve with zeal, punctuality and subordination?" (National Medical Conference).

At that time, the school and practice hospital area were close to the forest. The forest was home to many dangers, such as tigers, that made living and practicing medicine challenging for Professors Tôn Thất Tùng and Hồ Đắc Di and their respective families. Professor Tôn Thất Tùng actually forbade his children to wander far from the house for fear of tigers. Professor Hồ Đắc Di found it necessary to punt a floating raft across the Quẳng Stream to tend to his patients in Ải Village. There were only two professors at the school at the time. Professor Hồ Đắc Di wrote the following about the challenges of a physician, "Life, disease and death put forward such fearful matters to the physician every day. The desire to study and gain knowledge and the wish to cure people make me anxious because I felt terribly unhappy with my own dead end and awfully dumbfounded by the numerous mysteries of nature [...] If my limited professional qualifications sometimes make me unable to understand the deeper reasons for the pains that torture patients, culture allows me to relate on the grounds of common humanity, which is the also the purpose of true medicine."

This idea came naturally to Professor Hồ Đắc Di from his idea that a physician should be like a good mother. He believed that a physician's primary focus should be to cure patients; there were no other reasons for one's existence.

In the realm of university study and research, one matter that needed to be cleared up was the relationship between inheritance and development. Denying inheritance would lead to the tendency to fall into nothingness, but if inheritance was regarded unilaterally, it would be difficult to avoid conservatism and stagnation. Professor Hồ Đắc Di thought that knowledge was a continuous process that relies upon the lessons of previous generations. Thus he found it necessary to learn all the knowledge already accumulated in a particular field before trying to make new discoveries. In his words, "It is by getting on our precursors' shoulders that we can cast a glance beyond the wall which separates us from the unknown." So to insist on denying the eternal succession of the thing means to deliberately deny the evident truth.

However, Professor Hồ Đắc Di warned us not to be overconfident that what we have learnt as students consists of inviolable truths, because this approach can easily lead to severe opposition to new thoughts. Rather, according to Hồ Đắc Di, "We must study in doubt and realize in faith." It is necessary to recognize that successive scientific inventions - which can either supplement or negate each other - are just relative steps on the way to accessing absolute truth. The young generation still has to strive in the seemingly endless and hard process toward achieving that goal. Hồ Đắc Di's advice on the process was to, "Act as a thinker and think as a man of action."

As for the relationship between the individual and the organization in scientific research, Professor Hồ Đắc Di also contributed ideas unprecedented for his time. Apart from particular cases of genius, scientific research can only achieve great success gradually, by the consecutive contribution of research groups in many countries and from many related branches, just as a strong waterfall is formed from many individual droplets. However, innovative inventions are often the offspring of a single labor, a deep natural endowment of an inviolable personality. Breakthrough inventions like Albert Einstein's are the purest manifestation of individual intelligence and also the dialogue between thought and existence. Such a dialogue is often done by one man who has something unique – an internal genius that others lack. Such people are like eagles that, due to their abnormally large wingspans, can fly higher into the sky of basic scientific principles than others can.

Hồ Đắc Di wrote the following about such people, "They are exceptional men. But with them as the exception, scientific research generally requires an obstinate effort and patient work on the part of those who dedicate themselves to it. It earnestly asks for youthful qualities. Luck and attention lie in wait, but they only come with work, method, imagination." (*Mission and Role of the University*)

Scientific theories are sometimes contradictory and their longevity is often very short, just like a man's lifetime. What consoles the scientists are their real achievements and useful experiences that are succeeded by future generations in a limitless effort to enrich mankind's knowledge. The limited life of a scientist is integrated into the limitless life of scientific generations. As Professor Hồ Đắc Di wrote, "We are probably just the people digging the foundation for a sky-high castle coming in the next century."

No one can claim to know the future. It can be difficult to make rational decisions when confronted with contradictory scientific trends. However, over a half century ago, Professor Hồ Đắc Di affirmed judiciously that the soul of true medicine is humanism and that it is necessary to understand "the whole man as a creature with both a body and a soul." Man is a sensitive and reactive entity different from other lab animals. There are symptoms which can be found only in man. Humans bring happiness, misery, distress and anxiety into their diseases.

Humanism is the renaissance of ancient Greek principles. In medicine, it is the renaissance of Hippocratic and Pythagorean thought. In 1949, Professor Hồ Đắc Di commented as follows: "How strange it is when concepts of modern medicine return to concepts of ancient Chinese medicine."

Upon researching living organisms at super-micro levels, Professor Hồ Đắc Di also recognized that molecular biology and elementary particle physics had penetrated into medicine. He soon evaluated properly the increasing importance of social health and preventive sanitation. In addition, he had a judicious way of looking at Sigmund Freud's psychoanalytical theories even during the decades when Freud's theory was seen as heretical and anti-scientific in the Soviet Union.

After the liberation of Hà Nội from French occupation (October 10, 1954), Hồ Đắc Di also wrote massive theses on communism and medicine in the light of dialectical materialism, on the scientific-technological revolution, and on other subjects. However, he had already mentioned the most basic theoretical points between the years of 1946 and 1948, when he lived in the middle of the Viet Bac forest

Unfortunately, many of the articles that Professor Hồ Đắc Di wrote during the years he spent in Việt Bắc have been lost, and those that have been found have not been translated completely and thoroughly into Vietnamese.

Still Learning in Old Age

Throughout the years Hồ Đắc Di and his family continued to live in harmony with Professor Tôn Thất Tùng and his family. Though the scene was very different from the years in which they all lived in the forest amid tigers and leeches – not to mention the constant threat of military attack – the two families continued to live together in the peaceful years that followed the war. They lived together in a house with a beautiful garden near downtown Hà Nội, where they often walked together on the garden path or sat together at the dining table set out in their inner yard.

The two professors, who had been living together since they were 19 and 31 years old, were linked, not only by their common knowledge and experience, but also by their revolutionary fervor. In fact, Professor Tôn Thất Tùng continued to play an active role even after the fighting was over. On the occasion of World Environment Day, June 5, 1982, he publicly denounced the United States for its chemical war crimes in Việt Nam. Talking about his friend and colleague, Hồ Đắc Di said, "I have been living with Tùng since he was a 19-year-old student at Bưởi School. He is twelve years younger than me. Both us had to experience working in the French colonial hospital, which was like a prison without iron bars, but then we found the way to revolution in the light of Uncle Hồ's eyes. We both believed immediately and totally in his indulgent eyes and kind heart and then gradually realized the political ideal he was following. After that, through him and thanks to him, we understood the Party and communism."

Even when he was more than eighty years old, Professor Hồ Đắc Di kept his habit of reading. Every morning he would visit the library to read books and take notes. He would often get so absorbed in thinking that he would lose track of the time and everything around him. Aside from medicine, philosophy, literature, and the arts were his favorite fields to study. Inventions in nuclear and elementary particle physics, new ideas of genetics and molecular biology, and unique findings in information and system theories were not unfamiliar to this old man. Despite his age, he still maintained the same eagerness and thirst for knowledge that he had had as a resident student at the Medicine Faculty of Paris University more than 60 years before. He always modestly regarded himself as a "victim of a limited education level" or as a "lifelong student." He once said, "Despite my advanced age, I don't feel my soul aging. The fire of life in my heart is still warm and bright. I wish to convey to the younger generation my knowledge and love of life."

Toward the end of his life, Professor Hồ Đắc Di reflected on his decision to pursue medicine, "60 years ago, I was attracted to the humanism of medicine. I thought the infinite goodness and ardent love of life that a physician should have would warm my soul and bring me the passionate breath of life. And, among all careers, perhaps medicine and teaching are the two noblest careers as one brings life and the other brings knowledge. Both of them require a clean conscience. That was the reason for my choice at that time.

Looking back now, I feel like I love my career more because it connected two seemingly contradictory sides of me – science and morality. Medicine requires a thorough grasp of knowledge in many scientific fields such as surgery, physiology, chemical biology, chemistry, physics and mathematics. Yet, it also doesn't tolerate anything immoral because of its relation to human life.

If science and morality are combined cleverly, they will produce an intelligent sense like that seen in the arts.

How beautiful are the surgeons' hands when they skillfully do surgery on a human body – the most sophisticated and complicated among all creatures. How beautiful is the needle-work that heals cells and blood vessels to recover life. Isn't the cleverness of a surgeon the same as the talented brushwork of a famous artist?

However, just as in the ancient societies, the question is how to produce harmony between science and morality. Anyone who keeps that harmony within himself instead of sharing it with others is a fool!"

Up until he died in 1984, this revolutionary professor remained absorbed in the love of people and of life. When he was a alive, Professor Hồ Đắc Di said that one should "live like a teacher and die like a student," and he succeeded in living out this motto. Throughout his life, this professor, doctor, and scholar influenced many people and gained more knowledge than most; yet he remained a student who actively pursued knowledge even up to the very end of his life.

ĐỖ XUÂN HỢP – A MASTER OF ANATOMY

Since its foundation in 1931 during the French colonial period, the Hà Nội Institute of Anatomy has become a training school for many important doctors. Đỗ Xuân Hợp, Tôn Thất Tùng, Nguyễn Xuân Nguyên, Đinh Văn Thắng, Hoàng Đình Cầu, Phạm Biểu Tâm, and Nguyễn Hữu, among other people, all worked here as surgical assistants.

In 1932, the 26-year-old doctor Đỗ Xuân Hợp began to teach in the Anatomical Department of the Hà Nội Faculty of Medicine under the guidance of Professor Pierre Huard, Director of the Institute of Anatomy and Dean of the Medical School.

After ten years of studying and training, Hợp had amassed a valuable collection of documents on morphologic and anthropological anatomy of the people of Việt Nam and Indochina. In 1942, in collaboration with his teacher Pierre Huard, he published the book, *Morphologie humanie et anatomie artistique* (Human Morphology and Artistic Anatomy). This fundamental work in many disciplines – medicine, anthropology, archaeology, ethnography, and fine arts – had great resonance in both Việt Nam and France.

Seven years later, in 1949, his work was awarded the Testut Prize for international distinction in anatomy. At the time, ironically, Professor Đỗ Xuân Hợp was serving the resistance against French aggression. Together with his colleagues, he accepted the harsh conditions of life in the jungle and tried his best, with limited tools, to save wounded soldiers and to train army doctors and medical officers.

After the 1945 Revolution, despite difficulties, medical classes in Hà Nội still had sufficient amount of professional training materials, not to mention students were able to read textbooks in French. In the mountains during the heart of the war, it was necessary to find ways to surmount these difficulties as quickly as possible. Dr. Đỗ Xuân Hợp contributed to solving this problem by working countless hours to write lessons, translate texts from French, and coin new Vietnamese words to introduce unknown technical terms.

He recalled these exhilarating times in his memoirs, "There were nights when I couldn't close my eyes. Alone in a small hut in the black forest, beside a smoky oil lamp, I racked my brains to translate a medical term into Vietnamese, trying to find a simple and precise word."

The war produced more and more wounded soldiers. The percentage of amputations increased. Aware of this fact, he gathered his lessons and research to publish his 1952 book, *Giải phẫu tứ chi và thực dụng ngoại khoa* (Anatomy and Surgical Practice of Arms and Legs), a work with practical interest for training and for ambulances at the front. The doctor was congratulated by President Hồ Chí Minh and received the Order of Resistance.

Twenty years later, in 1971, he finished a complete series of textbooks on anatomy, with 1,500 pages and 900 illustrations. He also made considerable contributions to university-level medical training in Vietnamese by creating tens of thousands of new medical terms, avoiding Sino-Vietnamese words as much as possible. His treatment of the subject showed initiative, instead of routine academic procedures, combining systematic descriptions with localized anatomy and demonstrating applications for surgical operations.

In the preface to his work, he remarked, "The study of anatomy must serve surgical operation and pathology. Anatomy must serve the practice of internal medicine. That is why this work is written following a different concept from previous manuals. It has one section for localized description and one section for guiding operations. Thus students are not swallowed up in purely anatomical details and can remember the essentials for their professional activities."

This series, reprinted many times, was expanded in each new edition to add new experiences and to synthesize surgical practices from the two wars. Beside general concepts, he also described his discoveries on the morphology of Vietnamese people. Today, doctors who work in cities, the countryside, mountains, and remote areas still pay homage to his indispensable handbooks.

For thirty years since 1971, no other books on the same theme have surpassed his series of books. And, without a doubt, these are the most important among his twenty books and other works he did alone or in collaboration with others. The series comprises four volumes: *Giải phẫu đại cương và giải phẫu đầu-mặt-cổ* (General Anatomy and Anatomy of the Head, Face and Neck), *Giải phẫu tứ chi và thực dụng ngoại khoa* (Anatomy and Surgical Practice of Arms and Legs),

Giải phẫu ngực (Anatomy of the Chest), and *Giải phẫu bụng* (Anatomy of the Stomach).

Professor Đỗ Xuân Hợp (1906-1985) served as Director of the Military Medical Institute and advanced to the rank of Lieutenant General. Before his death at the age of seventy-nine, he was recognized as a Hero of the Armed Forces. He set an example as a conscientious doctor, a tireless researcher, a methodical educator, and a patriotic intellectual.

NGUYỄN XIỂN – A CONFUCIAN METEOROLOGIST

Throughout his entire life, which covered nearly all of the twentieth century (1907-1997), Nguyễn Xiển was an exemplary Vietnamese intellectual. He was born and trained under the French colonial regime and experienced trials and tribulations in the various national and international events. He had to make many decisions including whether or not to continue his studied based on the humanities or adopt the experimental science of the West. After receiving diplomas in France, he wondered if he should accept a lucrative mandarin post in the puppet royal administration in Huế or be satisfied with a modest high school teaching position.

In 1945, the Japanese put an end to French colonization through their March 9 coup, pretending to return independence to Việt Nam by creating a government of their choice. Nguyễn Xiển wondered whether he should join this government and wait for a more favorable situation or remained in his post as director of Phủ Liễn Observatory. After the August Revolution (1945) broke the French and Japanese yoke, he debated whether or not he should join the Hồ Chí Minh Provisional Government at the risk of being judged an opportunist. With the outbreak of the French war aggression in December 1946, he had to decide whether or not he wanted to give up the comfort of Hà Nội to serve the resistance in Việt Bắc. Finally, after two Indochinese wars, the American war having ended in 1975, the Government adopted the *đổi mới* (renovation) policy. He had to decide which attitude to take.

At each stage of his life, Nguyễn Xiển struggled with his conscience concerning his generation and the entire Vietnamese nation. Now that the century has ended, it seems in retrospect that solutions should have been obvious. In reality, these were veritable dilemmas at the time.

Most clear-sighted intellectuals cheerfully left their comfortable lives in the capital for the riskier adventure of resistance against French occupation. Nguyễn Xiển recalled the first street fighting in December 1946 in Hà Nội: "Barricades were erected on the pavement at each street corner. Great holes were made in the walls of houses to make escape routes. Residents evacuated the city in masses. My family went to Thanh Oai, in the adjacent province of Hà Đông, where my wife gave birth to a daughter. As for me, I remained in the city, constantly changing my residence. On the night of December 19, the war broke out. The enemy encircled my house, but failed to catch me. They killed the guard at my house. Dr. Nguyễn Văn Luyện and his son fought until death in their house.

The next morning, I still could take the last car to go to Sơn Tây to participate in the first meeting of the national Conference of Resistance and Administration. Then we had to go on moving by bicycle and on foot."

To either commit oneself openly for a cause or to lock oneself in an ivory tower away from conflict is a problem faced by many intellectuals. As a citizen of a colonized country, Nguyễn Xiển opted to struggle for national independence. This was a voluntary choice because he was born into a family of patriotic scholars, members of the anti-French Cần Vương (Serve the King) Movement. His brother fought for the Nghệ Tĩnh Soviet in 1930. When Nguyễn Xiển was a student in France, he had great admiration for Nguyễn Ái Quốc and knew young Vietnamese Communist activists such as Trần Văn Giàu and Phan Tư Nghĩa.

In August 1945, the Japanese surrender accelerated the course of events in Việt Nam. Nguyễn Xiển left Kiến An Observatory, where he had been director. He bicycled over 100 kilometers to Hà Nội just at the time when, at the appeal of the Việt Minh League, demonstrators encircled the Palace of the Royal Delegate of the North (under the Government created by the Japanese). The militia waited for the order to shoot at the crowd.

Nguyễn Xiển recounted, "I parked my bicycle, passed through the iron gate and asked to see the new Delegate Nguyễn Văn Chữ, who had replaced Phan Kế Toại. I told him, 'The Japanese have surrendered to the Allied forces. The Palace is now surrounded by the people. How can you resist them? To avoid a bloodbath, go and talk with them.' Chữ did what I told him while I rode back by bicycle.

Several days after that, Việt Minh officials Trần Đăng Ninh, Vũ Đình Huỳnh, and Phan Tư Nghĩa came to see me and asked me to serve the revolutionary government. This time, at the palace gate, I met the erudite scholar Nguyễn

Văn Tố, who told me smiling, 'Congratulations, dear sir! We are the first two Hà Nội intellectuals invited to sit in the cabinet. I've accepted the portfolio of Welfare and Social Affairs. I think you'd do well to accept Communications and Public Works.'

Then I met Võ Nguyên Giáp, who put forward the same proposal. I refused because I wanted to devote myself to technical affairs, and I suggested engineers Trần Đăng Khoa and Đặng Phúc Thông instead." In fact, Nguyễn Xiển did not want to be considered an opportunist if he accepted the proposal right away.

The next day, President Hồ Chí Minh received him in person and tried to dispel his reservations. "The state is in danger, President Hồ Chí Minh said. "No citizen should avoid his responsibility, as the old saying goes. You've received higher education. If you don't accept your responsibility, who will do it for you?"

"Mr. President, it's only that I'm not used to management."

"Nobody is used to it, sir. For the common cause, everyone does what he can."

Hồ Chí Minh's sincerity won over Nguyễn Xiển, who accepted his appointment as chairman of the Tonkin Administrative Committee, as well as the position of director of Phủ Liễn Observatory.

Nguyễn Xiển later confided, "I did everything that Uncle Hồ told me to do, not shrinking from any difficulties: repairing and building dykes, fighting against floods, forecasting the weather, teaching, researching popular science, the Fatherland Front, National Assembly, even foreign relations."

After the Declaration of Independence of Việt Nam (September 2, 1945), he proposed to President Hồ Chí Minh that the new government pay an official visit to the Hà Nội Temple of Literature in order to show its fidelity to humanistic traditions. His proposal was accepted. The delegation was led by the old patriotic scholar Huỳnh Thúc Kháng, Interior Minister.

He also took part in a 1946 government delegation to the festival of the Temple of Hùng Kings, the founders of Việt Nam. This was the first ancestral festival held after regaining national independence, although the trip was risky since the Phú Thọ area was infested by anti-Việt Minh reactionaries.

As chair of the Northern Administrative Committee in 1945, he had to fight day and night against floods that threatened thirteen provinces of the Red River Delta. Using his personal prestige, he succeeded in convincing engineers and private contractors to contribute generously to the strengthening of dykes.

All those who had the opportunity to know Nguyễn Xiển found him to be a fusion of two men: the scientist who is consumed by research and study (in meteorology, astronomy, mechanics, mathematics, and aeronautics), and the moralist advocating high Confucian humanistic virtues. Like many of his contemporaries from Nghệ Tĩnh, such as Đặng Thai Mai, Tạ Quang Bửu, and Hoàng Xuân Hãn, he also synthesized ideas from the East and West. He was given the Hồ Chí Minh Award for his two books: *Đặc điểm khí hậu miền Bắc Việt Nam* (Characteristics of the Climate of Northern Việt Nam) and *Tập bản đồ khí hậu miền Bắc Việt Nam* (Climatic Maps of Northern Việt Nam).

NGUYỄN VĂN HUYÊN – A BRILLIANT VIETNAMOLOGIST

Growing up in Hà Nội and educated in Paris, Nguyễn Văn Huyên became one of the first few Vietnamese to defend his Ph.D. thesis in the field of social sciences and humanities in France. It was published and had a widespread reputation in Europe.

On coming back to Hà Nội, he worked at École Française d'Extrême Orient (French School of the Far East). He continued to contribute many profound research works on Vietnamology.

Being sincerely patriotic, he participated in the Vietnamese Revolution and Resistance. He was entrusted with the great responsibility of serving as Minister of Education by President Hồ Chí Minh for twenty-nine years until he passed away.

Hard Childhood

Nguyễn Văn Huyên was born on 16 November 1908 in Thuốc Bắc Street, Hà Nội. His hometown was in Lai Xá Village, Kim Chung Commune, Hoài Đức District (present-day Hà Nội). At that time, Thuốc Bắc Street was known as Hàng Áo Street. When he was eight, his father, Nguyễn Văn Vượng - an officer at Hà Nội Treasury Department, passed away. His mother, Phạm Thị Tý, raised her children on her own. She worked as a needlewoman. Every day, she packaged into cloth bags, put them on her shoulders, and sold them at the market. She was a very clever tailor. She always changed designs according to customers' predilection. She eventually earned enough money to buy her own shop at 30 Thuốc Bắc Street.

When Nguyễn Văn Huyên was a child, his mother sent him to school to learn Chinese characters with the hope that he would carry on his grandfather's career of making oriental medicine. Later, his mother was reluctant to move him, his sister, Nguyễn Thị Mão, and his younger brother, Nguyễn Văn Hưởng, to a Western school. However, the few years of learning Chinese characters were not completely useless to that industrious boy. Afterwards, when doing research work, he often adequately annotated Chinese characters, Vietnamese Nôm script, and related Tày Nôm script to conveniently research and ensure high accuracy. At the Western school, Nguyễn Văn Huyên was still a good student. He was given a reward at Hà Nội Opera House every year.

An Industrious Youth

After passing the first high school finals, Nguyễn Văn Huyên was sent to France on 2 December 1926 to continue studying with his younger brother, Nguyễn Văn Hưởng. After passing high-school finals in France, he applied to Montpellier University, a long-standing French university established in 1221. At that time, Vietnamese students in France gathered in Aix-en-Provence and Montpellier, which are endowed with a warm climate. They were mostly children of plantation owners in the South. A few were descended from Confucian scholars and officials' families in the northern and central parts of Việt Nam. Montpellier University was divided into three separate colleges. Montpellier 3 received foreign students studying the French language and literature. It was also named after Paul Valéry (1871-1945) – a famous French writer and scholar who had strong influence on Nguyễn Văn Huyên as well as Tôn Thất Tùng, a world-famous Vietnamese surgeon.

Huyên and Hưởng were very resourceful and economical. During the winter, not having fashionable coats, they wore old sewn cotton coats which were sent by their mother. In the evening, they ate in a small restaurant, usually eating the same dish every day. On days off, each of them rode a bicycle to pick grapefruits with farmers. They were able to eat grapefruits freely but received no wage. Once, they rode to Mount Aigoual on the Massif Central mountain range in the central part of France. By 4:00 p.m. the snow was up to their knees, but they still had to bike up. Near midnight, they arrived at the guest-house. Luckily, they were not frozen from the cold. Climbing and going to picnics were Huyên's favorite activities because they were not as expensive as going to the swimming pool. Perhaps the frequent exercise prepared Minister Nguyễn Văn Huyên for riding from the northernmost area of Việt Nam to Military Zone 4 across a 200-km road through many mountain passes and slopes during the Anti-French Resistance War (1945-1954).

When studying in France, Nguyễn Văn Huyên communicated with intellectuals such as Nguyễn Mạnh Tường, Phạm Đình Ái, and Nguyễn Tường Tam (Nhất Linh, writer). Afterwards, Professor Nguyễn Mạnh Tường stated: "Huyên was more agile than I was. I had not thought about where to go during the summer holiday; he anticipated everything. He invited me to travel around Europe instead of staying only in France, going to one place each year. During the first year, we went to Spain, then Italy. Upon returning, he said, 'Traveling forms a young man, we cannot learn anything when we just stay at home and dance with ladies!'"

In the second year, he said, "We are studying Greek and Roman ancient literature and history, but it is really a pity that we have never set foot in Athens or Rome!"

In France, there was an association of people who researched Greek and Roman ancient literature and history. Every year, they rented a ship from Marseille to travel around the Mediterranean Sea and visit cities along the shore. The atmosphere on the ship was very interesting because the members of this association were mostly scholars or at least bachelors. The lifestyle on the ship was not busy and hurried, but easygoing. You could lean on the banister on the deck, enjoy the view and meditate. Through economical voyages, we could set foot in Venice, Milan, Vienna, Geneva, and Budapest, and stop in Athens and Istanbul (Turkey), which is situated on the strait between Europe and Asia.

Additionally, during the years that Nguyễn Văn Huyên and Nguyễn Văn Hưởng studied in France, their sister Nguyễn Thị Mão helped them a lot. After passing the final exam at Félix Faure, an all-female high school, Nguyễn Thị Mão became the first Vietnamese woman to enter the Indochina College of Education. Afterwards, she taught mathematics at Đồng Khánh All-Female Secondary School (present-day Trưng Vương Secondary School, Hà Nội). Her salary was very high, about 120 piastres. Each quintal of rice cost only a few piastres. However, every month she kept only 20 piastres for herself. The rest, 100 piastres, was contributed, with her mother, in buying checks to send to France in order to help her brothers.

"A Great Event in the History of Sorbonne"

On 17 February 1934, Nguyễn Văn Huyên successfully defended his Ph.D. thesis at Université de Sorbonne (Sorbonne University). His main dissertation was entitled *Alternative Singing of Vietnamese Youngsters,* and his auxiliary dissertation was *Introduction to the Study of Southeast Asian Stilt-houses.* Chairman of the Dissertation Jury Vendryès considered them "a significant event in the history of Sorbonne." Highly appreciating the value of these two projects, Director of Paris Academy Sébastien Charléty examined and agreed to publish the works. Immediately after that, Paul Gauthner Orient Publisher printed Nguyễn Văn Huyên's two dissertations as books. These books received much critical acclaim. Many specialist magazines in France, Germany, the Netherlands, and other countries almost immediately published commentaries on those books.

Showing his adoration of Paul Valéry, Nguyễn Văn Huyên used a sentence in a poem by this French poet to put in the preface of his main dissertation, "Le vent se lève!... Il fault tenter de vivre! (The wind is rising!... We must try to live!)"

The dissertation in French is entitled *Les Chants Alternés des Garçon et des Filles en Annam* (Alternative Singing of Vietnamese Youngsters).

The dissertation was published in Paris in 1934. However, it wasn't until 1995, 61 years later, that it was translated into Vietnamese. There were limited copies, so many Vietnamese readers did not know about these works, including researchers in social sciences and humanities in Việt Nam. Many of these researchers were educated after 1945 and were not fluent in French. Therefore, they could not understand it thoroughly, let alone ponder on, inquire into, or criticize it. Consequently, few Vietnamese scholars could then go further on Nguyễn Văn Huyên's path – that is, the path he paved by applying modern international academic knowledge to the clarification of historical and cultural issues in Việt Nam through serious, internationally standard, and scientifically persuasive studies on Vietnamology.

It is ironic that such late-known research works in Việt Nam have been approved to be honored in large cultural centers all over the world. On 20 March 1934, in a letter for Nguyễn Văn Huyên, the French poet and scholar Paul Valéry wrote,

"Dear Mr. Huyên,

Thank you for writing to me and sending me your research! I am grateful to you for putting a sentence from my poem "Le Cimetière Marin" (The Graveyard by the Sea) in your preface. I have become more sensitive with the method you chose in working out the composition of poetry in Vietnamese festivals. The harmony of sounds and senses is the primary goal in poetry, but this goal is not specified in our languages. It can be obtained only by facts, and it seems to be a contingency as they can never be sure to bring about the musical emotional endurance for mentality.

Nevertheless, I found in your book examples of poetry forming in the developing state. I found there the states of songs, and poems from songs and rhythms. And I suddenly remember Ronsard, who composed poems with a lute. I also remember myself composing many poems derived from the melodic images that suddenly would come and obsess me; those images gradually became words, and then an idea at last.

Everything you mentioned about symmetry, balance, and groups made me extremely interested. (However, I do not understand Vietnamese, so I can only yield superficially from your analysis.)

Thanks to what you wrote about Vietnamese poets, I am pleased to read what I believe to be true with all poetry, but in our country these observations are rarely known or understood. When I ventured to mention this, some thought that it was offensive; some thought I was too sensitive and complicated. Yet, I felt that I did pay attention to the origin.

Please believe in my appreciation and sincerity!

Paul Valéry"

We believe that this is not simply a letter of etiquette, but a brief and profound assessment of the main dissertation of Nguyễn Văn Huyên, which is partly in accordance with Paul Valéry's concepts of poetry. Today, we can easily find the Vietnamese translated versions of the 160-page dissertation in 24x16 cm size paper. The second volume of *Nguyễn Văn Huyên toàn tập* (Complete Works of Nguyễn Văn Huyên) was printed in July 2001 by Việt Nam Education Publishing House. Additionally, at the beginning of 1934, when Nguyễn Văn Huyên completed his thesis, he was only twenty-six while Paul Valéry was sixty-three.

A Resounding Reputation

Not long after that, Professor Christoph Furer Haimendorf, a well-known ethnologist, wrote a summary of Nguyễn Văn Huyên's project in *Anthropos*, a world-famous magazine in German. The summary, according to Professor Hà Văn Tấn, was followed by the following comments, "There is no doubt that seasonal festivals, which include alternative singing, dancing, and flirting, belong to the agricultural civilizations in Indochina and South China.... In terms of their characteristics, these festivals exhibit bilateral freedom, which also means women's freedom in choosing their fiancés. Therefore, these festivals originate from civilizations with strong nuances in matriarchy." (30.5-6 [1935])

In *Orientalische Literature Zeitung* (Oriental Literature Magazine), H. Jensen wrote, "This work, whose author is a Vietnamese who used to give lectures at École des Langues Orientales (Oriental Languages College) in Paris, is a valuable contribution to understanding Việt Nam. It covers two fields. On the one hand, it provides ethnologists and folklorists with rich materials that notice all details in understanding love, first in Vietnamese courting and marriage customs. On the other hand, it also contains many insightful things for linguists.... The author is right in emphasizing that in daily conversations, Vietnamese uses fixed phrases, also called standardised forms, so they can easily compose poems. This book consists of many of such model parts and extemporizations. These model parts originate from profound collections." (11 [1935])

However, according to Professor Hà Văn Tấn (a scholar who received the Hồ Chí Minh Prize), Jean Pryluski appreciated Nguyễn Văn Huyên's work the most. In a long article entitled "Rhythm and Poetic Extemporization (Regarding a Newly-published Book)" posted in *Journal de Psychologie Normale et Pathologique* (Normal Psychology and Pathology Magazine), he wrote, "In the West, language is commonly different from poetry; language became an analytical tool, and to some people it tends to be similar to the close accuracy of scientific symbols. In Việt Nam, the distance between words and lines of poetry is not great like that. Vietnamese people often make their words rhythmic, and in every case, they combine their words to symmetric groups which do not need to be modified, but become elements of poetry.... After noticing that Vietnamese poetry is formed by symmetric groups of words, Nguyễn Văn Huyên studied the structure of these symmetric groups and how to combine them." (9-10 [1934])

As mentioned above, after defending his main dissertation, Nguyễn Văn Huyên also defended his auxiliary thesis entitled *Introduction à l'étude de l'habitation sur pilotis dans l'Asie du Sud-Est* (Introduction to the Study of Southeast Asian Stilt-houses). This thesis was also published and received a hearty welcome from the public. It is widely known that in many major universities, there are a large number of postgraduates that defend their Ph.D. theses every year. Very few of theses are published, and not every printed book receives wide recognition. Therefore, it is evident that there is significant value to Nguyễn Văn Huyên's work. According to Professor Vendryès, the chairman of the dissertation jury, this was "a significant event in the history of Sorbonne."

In the book review in *Anthropos*, Rudolf Rahman wrote, "Apart from written documents, the author also exploited the collections at European museums and distributed his ready-printed questionnaires. Therefore, the author could provide an invaluable overview of stilt-houses. Other scholars will surely agree upon Nguyễn Văn Huyên's opinion that their contemporary popular form results from not only natural conditions (such as the ground's structure and climate) but also external factors like borrowing and inheritance." (30.5-6 [1934])

Scholar Jules Sion published a long comment on Nguyễn Văn Huyên's book with the title "A New Type of Residence: Stilt-houses" in *Annales d'histoire économique et sociale* (Annuals of Economical and Social History). He wrote, "In contrast to the long-term surveys by geographers in rural residences in Europe, here, in a different layout, is an earnest book written about the predominant type of residence in Southeast Asia. The author described different forms varying from one nation to another, according to tourists as well as the restorations in Italian or Roman museums. Then, he drew the common characteristics. He applied geographical methods to study the reasons of the distribution, cultural events attached to this type of residence, etc. Thanks to him, we can glimpse ten problems in European civilised relations. In order to explain, he proposed a useful list in which different forms and characteristics of this type of residence can be found." (9: 499-501)

Modern Scientific Research Method

Now reading Nguyễn Văn Huyên's book again, I am surprised and admire the interesting profound content as well as the modern scientific research methods. His references include 103 authors of different nationalities.

These nationalities include French, English, American, German, Dutch, Austrian, and more. At that time, Indochina was the land located between India and China and included Myanmar, Assam, Thailand, Laos, Cambodia and Vietnam. Nguyễn Văn Huyên surveyed the Palaung people's stilt-houses in the north, the Intha people's in the south, the Wa people's in the east, and the Chin people's in the west of Myanmar. After surveying the stilt-houses of Thailand, Laos, and Cambodia, he spent a lot of his time and effort in surveying the stilt-houses of different ethnic groups in Việt Nam. He especially described thoroughly different forms of stilt-houses of different groups of Dao people such as Dao Cốc, Dao Cao Lan, Dao Chàm, Dao Đeo Tiền, and Dao Ngọc Điền. He also surveyed Thượng people's stilt-houses in different places of Việt Nam such as the Trường Sơn mountain range, Đác Lắc Highland, Bình Định, and Quảng Ngãi. Obviously, he did not forget to describe the stilt-houses of populous minority ethnic groups in Việt Nam such as Tày-Nùng and Mường. Southeast Asia also included Indonesia. Therefore, Nguyễn Văn Huyên studied stilt-houses of ethnic groups residing in the Sumatra Islands such as Atjeh, Gajo, Karo Batak, Toba Batak, Minangkabau, Lampong, and Benkoelen. Next, he continued surveying other ethnic groups' stilt-houses on Bornéo Island, Célèbes Island, and small islands such as Halmahera, Boeroe, Ceram, Wetar, Timor, Tanimbar, and others.

To collect materials for his auxiliary dissertation, the author worked at the Ethnographic Museum in Trocadero, the Natural History National Museum in Paris, and the Office of the Indochinese Economy. He also went to the Netherlands twice. The first time he went was at the end of the summer in 1931. The second time, he went in early 1932 and worked at Leyde Museum, Rotterdam Museum, and Amsterdam Institute of Colony where there was a museum which allowed him to explore many materials related to the stilt-houses in Indonesia – a Dutch colony at that time. He met and interviewed people returning from Indonesia in order to have a better understanding of the field that he was studying. Nguyễn Văn Huyên also went to Rome and worked at Latran Museum, and used collections that Catholic clergymen had presented Vatican through the sponsorship of Reverend Ladeuze.

Through Nguyễn Văn Huyên's auxiliary thesis as well as his other research works, we find that he never hastily generalized or drew arbitrary conclusions based on prejudice. If information was not accurate or adequate, it was not included.

Nevertheless, Nguyễn Văn Huyên also doesn't belong to the type of scholars who are easily confused by so many materials that they are unable to reach any generalization. The diversified data in the part of descriptive monograph set a solid base for conclusions made in the part of general essay. In this part, the author concentrated on explaining the reasons for the stilt-houses' existence.

The Friendship of Two Young Ph.D. degree holders

Nguyễn Văn Huyên and Nguyễn MạnhTường both defended their Ph.D. theses around the same time. Nguyễn Mạnh Tường then came back to Hà Nội. Nguyễn Văn Huyên still stayed in France for a short time because he had to continue with his teaching contract at École des Langues Orientales. Before saying good-bye, Nguyễn Mạnh Tường told Nguyễn Văn Huyên, "We are both new Ph.D.

degree holders, we need Ph.D. uniforms to wear! (The Ph.D. uniform is different from the graduate uniform and M.A. uniform. A Ph.D. uniform has three rows of rabbit's fur while the teacher's uniform in college has only one). I don't have enough money to go to the tailor's. You have been teaching and have a salary. Can you help me?"

Nguyễn Văn Huyên smiled and replied, "I have, why don't you?"

Then, they went to the tailor's to have the Ph.D. uniforms made. Yet, according to members in Professor Nguyễn Văn Huyên's family, he never wore it because he liked simplicity and wished to remain incognito.

While still studying in France, they agreed that when they returned to Việt Nam, they would not work for the government. They would just teach, study, and write books. Therefore, in 1935, they both taught at Lycée du Protectorat (also known as Bưởi High School). Nguyễn Văn Huyên taught history and geography while Nguyễn Mạnh Tường taught literature. At that time, only Bưởi High School had a teaching staff with such high qualifications. According to Professor Nguyễn Mạnh Tường, "the Governor-General and Resident Superior extremely respected" the school. However, everything did not live up to their expectations. Native teachers were not allowed to teach certain subjects. Nguyễn Văn Huyên decided to move to École Française d'Extrême Orient. Nguyễn Mạnh Tường established a law office.

Continuing Studies in Vietnamology

In 1938, Nguyễn Văn Huyên was appointed as a permanent member of École Française d'Extrême Orient, and in 1941, as a permanent member in the Indochinese Scientific Research Council. Highly appreciative, Georges Coedes, the rector of the French School of the Far East wrote in the preface of Nguyễn Văn Huyên's book entitled *Le Culte des Immortels en Annam* (God Worship in Việt Nam) (1944), "Along with the firm education he had at the Faculty of Literature and Faculty of Law in Paris University, France, Nguyễn Văn Huyên also had the unique advantage of grasping Việt Nam's social events through personal and innate experiences, having an ability of approaching sociological materials directly and immediately, and going into details to carry out surveys which European scholars could not do. These trump-cards helped him conduct and successfully finish many research works on the religious, spiritual, and social life of rural residents in Việt Nam." (5)

During ten years (1934-1945), Nguyễn Văn Huyên made his forty-six studies public. These were mainly in French at first and were translated into Vietnamese later. The first and second volumes of the more-than-2000-page collection of *Nguyễn Văn Huyên toàn tập*, a collection that every Vietnamese intellectual should have in his family's bookshelf, were printed in 2000. Vietnamese people who want to understand their origins, beliefs, folk festivals, national culture and civilisation can find many useful and reliable things in Nguyễn Văn Huyên's studies. In a study on the worship of tutelary gods entitled *La carte de repartition de génies tutélaire*, he wrote, "Based on the time villages were set up, there is no doubt that the worship of tutelary gods is one of the solid elements of Vietnamese civilisation." (23)

Nguyễn Văn Huyên paid particular attention to immortal deities such as Tản Viên, Gióng, Chử Đồng Tử, and others.

Immortals are worshipped in many places, especially in Vĩnh Phúc, Phú Thọ, Hà Nội, Hòa Bình, Hà Nam, Nam Định, and Ninh Bình. He commented, "Studying the worship of these gods, we can find the role of Vietnamese national history in Southeast Asia. This nation did not choose to copy China, but created their distinct life themselves which has always been kept youthful throughout history."

While researching, he silently expressed his patriotism. While some Vietnamese intellectuals became "Westernized," dismissing the beliefs and folk cultures of Kinh and Tày peoples, he showed his appreciation by studying them. He studied the Phù Đổng Festival and Lý Phục Man, a tutelary god in Yên Sở Village on the outskirts of Hà Nội. He then created a map of the distribution of tutelary gods in Bắc Ninh Province. He studied the birth of Nội Đạo Tràng in Việt Nam, the legend of Phạm Viên, a god in Việt Nam, as well as the worship of many other Vietnamese gods. Georges Coedes also wrote in the preface of *Le Culte des Immortels en Annam* (op. cit.), "The gods, mentioned here, who lived on earth with temples and divination rituals are purely Vietnamese gods. These are national gods whose brilliantly beautiful legends are an inseparable part of the religious and spiritual heritage of the country. Europeans will discover one aspect of this religion that they have had little understanding of before, thanks to this book. So although this book does not have literary ambitions, it still gives out a subtle and secret aroma, the fine poeticism of Vietnamese legends." (vii)

The project entitled *Songs in Tày people's weddings in Lạng Sơn and Cao Bằng* helped introduce readers to Tày Nôm script as well as hundred of songs written in Tày Nôm script and translated into Vietnamese. It was a 320-page research work which reflected the author's diligent efforts. Mrs. Vi Kim Ngọc, Professor Nguyễn Văn Huyên's wife, was a Tày person from Lộc Bình District, Lạng Sơn Province. This probably encouraged and enabled him to learn about Lạng Sơn, Tày people's culture, and the beautiful lyrical songs there.

Văn minh Việt Nam – An Excellent Work of Research

In 1938, Nguyễn Văn Huyên helped Hà Nội Law University establish the Department of History of the Vietnamese Civilisation. The manuscript of the book *Văn minh Việt Nam* (Vietnamese Civilisation) was completed in 1939 and was used for teaching. It wasn't published until 1944. The reason for the delay was that the manuscript had to wait for the French authorities' approval. About this book, Professor Phan Hữu Dật wrote, "Today, reading this work, we can not help but be surprised at his wide knowledge. It is not exaggerative to say that the research work *Văn minh Việt Nam* not only successfully summarizes his research work in ethnology during ten years of working diligently and creatively but also wraps up the achievements of scientific branches studying Vietnamese people until the mid-twentieth century.... Issues of homes, villages, and the country are equally vital at present in the building of the political system of the country during the renovation."

The 270-page Vietnamese version of the book *Văn minh Việt Nam*, which was printed in the first volume of *Nguyễn Văn Huyên toàn tập*, published in 2000, deserves to be read carefully.

As I mentioned above, Nguyễn Văn Huyên respects the native uniqueness of national traditional culture. Yet, that does not mean that he was extremely xenophobic or denied the positive influence of Chinese culture on Vietnamese culture. In the book *Văn minh Việt Nam*, Nguyễn Văn Huyên wrote, "The purpose of Confucianism is governing people effectively." He quoted a Confucius's saying, "If the emperor is solemnly close to his people, they will respect him; if he is grateful to his parents and is kind-hearted, his people will be loyal to him; if he supports good men and educates bad men, they both will try their best." (254)

This means that the "emperor must act like an emperor." The person who holds the supremely powerful position must deserve this position. In terms of the morality of the emperor, Nguyễn Văn Huyên specified, "The very first thing he must do is study literature, read the books written by the ancient sages, and learn about civility and music. Moreover, he has to improve his self to reach moral perfection and grasp the best virtue of Confucianism which is benevolence, the love for others." (259)

Nguyễn Văn Huyên believed that the subject of ancient Orient should be taught compulsorily at high schools. He wrote, "Vietnamese had to wait until the nineteenth century to become a civilised language because Chinese had been the only official scholarly language in Việt Nam for nearly 2000 years.... Chinese played an outstanding role in Việt Nam. They left unforgettable vestiges in Vietnamese institutions as well as intellectual expression.... This is easy to understand. This country (Việt Nam) has had frequent relations with China, which has a highly developed civilisation and many works of literature."

Nguyễn Văn Huyên was aware of the important role of "Confucian scholars," the feudal intellectuals in Việt Nam's ancient society. They not only mastered the science of defending the country and opposing invaders but also took the noble responsibility of the teachers - "soul engineers" - for the entire population. Unfortunately, his free and objective point of view was not widely known for a long time until it was eventually developed as a specific educational policy.

Participating in the Revolution and Resistance

The August Revolution in 1945 created a turning point in the life of this patriotic and scholarly intellectual. Once, he said, "After many years, I still remember my feelings when standing at the historic Ba Đình Square and listening to President Hồ Chí Minh read the Declaration of Independence on 2 September 1945. It was not the first time I had heard the words 'independence and freedom', but at this moment, and through the Resistance, I perceived the deep meanings of these two words."

A few days earlier, on the afternoon of 22 August 1945, four famous intellectuals in Hà Nội, namely Nguyễn Xiển, Nguyễn Văn Huyên, Ngụy Như Kon Tum, and Hồ Hữu Tường, sent a telegram to Emperor Bảo Đại. The telegram said, "A provisional people's government has been established. Hồ Chí Minh is the president. You are requested to abdicate to reinforce and unite our country's independence."

Since 1938, Professor Nguyễn Văn Huyên had participated in the Management Board of the National Language Propagation Association. After the August Revolution in 1945, he was appointed Director of the Higher

Education Department, Ministry of National Education, by Hồ Chí Minh and Rector of the French School of the Far East.

On the morning of 15 November 1945 in Hà Nội, Việt Nam University held a ceremony to begin its first academic year. This university included five faculties, namely Medicine, Sciences, Literature, Socio-politics and Fine Arts. President Hồ Chi Minh and many governmental officials attended the ceremony. Professor Nguyễn Văn Huyên, the director of the Higher Education Department then, delivered a speech to specify, "In this ceremony, we, professors and students, want to show the world that in this solemn moment of our country's future, in addition to our efforts in the struggle at the front, our nation also makes attempts to take part in developing the human culture. We also hope that this new higher education will be a strong force in our national combat forces. We want it to become the rampart so that the protracted resistance can completely restore our territory, liberate the spirit of our nation which has more than four thousand years of independence and has established a unique civilisation along the Pacific coast ourselves.... This national university is opened in difficult contexts. In the world, the war has been stamped out for over a month. A massacre has broken out in the vast and most fertile strip of land in the south of this country. Based on unfair and deceitful theories, French colonialists were against the human cause of freedom. The invasion has been continuing fiercely, so we are commanded to build up the basis of the National University."

On the morning of 9 February 1946 at the National Museum in Hà Nội, the exhibition of national history and culture organized by Professor Nguyễn Văn Huyên and officials from the French School of the Far East was inaugurated. President Hồ Chí Minh and many government officials attended this exhibition. It was the first time visitors could witness many precious materials and exhibits which they had only heard of or seen in photos in specialized magazines. Professor Nguyễn Văn Huyên also participated in two historic conferences related to the destiny of the country: Đà Lạt Conference and Fontainebleau Conference.

In November 1946, President Hồ Chí Minh met privately with Professor Nguyễn Văn Huyên and persuaded him to fill the role of Minister of National Education because Hồ Chí Minh believed that he had both talent and virtue. Certainly he would have difficulties in his task, but Hồ Chí Minh and the union would help him wholeheartedly. Hồ Chí Minh said briefly, "The reactionaries betraying our country showed their true faces and left. French colonialists are very bull-headed and warlike. However, they can not do anything because with our whole heart we will surely win." His belief was extremely vehement.

Professor Nguyễn Văn Huyên held office as Minister of Education, beginning in November 1946. He held this office for twenty-nine years throughout two resistance wars against France and America until he passed away on 19 October 1975.

During nearly three decades, he made great contributions to the building of the modern educational system in Việt Nam. Nevertheless, because of being busy with directing processes related to education, he did not have enough time to write new unique works on Vietnamology. In 2000, the collection of Nguyễn Văn Huyên's research works was awarded the Hồ Chí Minh Award in social sciences and humanity.

PHẠM NGỌC THẠCH – A PHYSICIAN AND REVOLUTIONIST

Over thirty years ago, on November 7, 1969, Dr. Phạm Ngọc Thạch died from biliary peritonitis and pernicious malaria in the Tây Ninh resistance base of southern Việt Nam. At 2:00 a.m. that day, shivering with fever, he had still operated in the pale light of an oil lamp in a cave that served as hospital for the front. Yielding to his students' insistence, he had gone to sleep. However, he never woke up.

When the news of his sudden death came to Hà Nội, Professor Hồ Đắc Di, Rector of the Hà Nội Medical University and President of the Việt Nam Association of Medicine, expressed the pain of all his colleagues: "Dr. Phạm Ngọc Thạch is no longer in this world! It is not possible! But I must accept reality. When I read the obituary in newspapers, my eyes blur with tears. He always seems to be present, standing here, sitting there... How can I forget his image and his voice?

It has not been so long since I saw him. I still remember our family meal before his departure for the south via the Hồ Chí Minh trail. Tôn Thất Tùng and Trần Hữu Tước were with us. He discussed subjects for medical research and urgent problems of public health in the southern resistance. He promised to continue our discussions after his study trip in the south. I have the impression that the heat from our last handshake still remains in my flesh."

Upset by his death, sculptor Diệp Minh Châu worked relentlessly for seventy-two hours to carve Phạm Ngọc Thạch's bust for his funeral. In his message of condolences, Jean Paul-Sartre expressed his admiration for the generosity, intelligence, and devotion of the deceased man. The eminent mathematician Laurent Schwartz was profoundly distressed by the passing of a "great physician and statesman."

At the time of the French colonization, before the August 1945 Revolution, Dr. Phạm Ngọc Thạch had opened a clinic in Sài Gòn. The physician was known for his sincerity and his kindness, not to mention his competence. Under French domination, the number of private doctors in each district could be counted on one hand. Only rich people could afford to consult them.

One night, a poor old woman who lived in a squalid hut in the suburb was stricken with a serious illness. Her son, not knowing what to do, came to Dr. Phạm Ngọc Thạch's villa. He tried knocking on the door. Through the gate, he saw a half-naked young man in the garden emerging from a mosquito net hanging over a folding bed. He thought he had woken up a servant. After a minute, he realized that it was Phạm Ngọc Thạch himself. Dr. Phạm Ngọc Thach then drove to the patient's cottage.

All Sài Gòn knew that Phạm Ngọc Thạch was the physician of the people. He benevolently took care of workers, coolies, rickshaw pullers, dockworkers, and newspaper vendors.

After brilliant results studying medicine in Việt Nam and France, he became a respected specialist on tuberculosis and was the only Indochinese member of a French tuberculosis research society.

His consulting room received not only poor people but also revolutionary militants who were ill and needed medical attention while still in prison. Before the Japanese overthrew the French colonial administration in March 1945, he had organized the Thanh niên Tiền phong (Vanguard Youth) movement in the south to act for national liberation. Its members joined the August 1945 Revolution. After this historic event, he was appointed to the Provisional Revolutionary People's Committee of Sài Gòn.

Professor Hồ Đắc Di recounted, "In 1931, when I began to teach at the Hà Nội Faculty of Medicine, I heard of a certain Tư Thạch, a sixth-year student who was going to defend his doctorial thesis. His enthusiasm, his intelligence, and his passion for study and research made me think that he would be a highly motivated young intellectual and a brilliant physician in the future.

A long time later, in 1953, amidst the preparations for Điện Biên Phủ campaign, I met him again in Việt Bắc, the revolutionary resistance base led by Hồ Chí Minh. He came from the South to receive a new mission. On his way to the central authorities, he came to see me and Tôn Thất Tùng. I was very ill, thin, and suffering from pernicious malaria. He asked to stay with me for a week to take care of me. He was constantly by my bedside. I was deeply touched by his conscience as a revolutionary physician and his devotion to a friend and colleague. I owe him my life. I was grateful to the revolution for having given him such feelings.

Despite being an intellectual privileged by the old regime, he has renounced everything to take sides with the revolution, connecting science and revolution. I've always followed his steps."

In 1954, when the northern half of the country was liberated, he was appointed Minister of Health of the Democratic Republic of Việt Nam and Director of the National Institute of Tuberculosis.

Though involved in numerous political activities, he did not neglect medical science. He led a group of pulmonary specialists who, after three years of hard work, succeeded in producing the BCG vaccine. This vaccine, which could be stored for a long time without refrigerators, was easily to transport from rice fields to mountains. It resulted in a significant reduction in the number of patients. Over sixty research institutes in forty countries have asked Dr. Phạm Ngọc Thạch to send them documents on the BCG vaccine. *Tubercle*, a British journal of the Research Society on Pulmonary Diseases and Tuberculosis, ran an article on the vaccine with the following comment, "For developing countries, the preventative anti-tuberculosis method of Dr. Phạm Ngọc Thạch is the best."

On the basis of his rich clinical experiences and his vast learning, he widened his research to include different pulmonary diseases: lung polyps, lung parasites, dry hemorrhage bronchiectasia, anthracosis, silicosis, and so on. His pioneering work opened a new chapter for lung treatment in Việt Nam. He was the first to study the injection of biostimulin of Filatov into jing-luo (energy pathways in traditional medicine practice) in combination with antibiotics for the treatment of phthisis. Tens of thousands of pulmonary tuberculosis patients have been saved in this way. This initiative of Vietnamese medicine has provided a solution to the difficult problem of fighting against tuberculosis in poor countries: how to produce effective, simple and cheap treatment?

As a revolutionary physician, he fought particularly hard against the regressive medical legacy of colonialism and feudalism. He paid special attention to the prevention and elimination of social diseases. As Minister of Health, he presided over the training of an important number of health officials and medical researchers. Prime Minister Phạm Văn Đồng paid tribute to "this physician who, by his conception, his thought and his way of research, has found original and creative solutions to difficult and complex problems concerning the elimination of diseases inherited from the old regime."

Professor Hồ Đắc Di went on, "I more than once discussed with Minister Phạm Ngọc Thạch about the future of Vietnamese medicine in the perspective of the world medicine. The depth of his theoretical and practical views impressed me. He has shown a great attachment to workers, peasants, soldiers, and especially to our southern compatriots in the occupied half of the country. During one of our trips abroad, I remarked that he didn't get tired of learning what was new, this didn't hinder him from keeping his personal opinions. We had, he said, to grasp well the realities of our country and to show creativity. Well-known scientists whom I've met have expressed their admiration for his national pride and the independence of his spirit."

TẠ QUANG BỬU – AN IMPARTIAL AND INTELLECTUAL TEACHER

Out of Nam Đàn – A Land of Learning

Ever since I was a child, the Western-educated Tạ Quang Bửu has been an excellent role model, a far-away but accessible legend.

Professor Tạ Quang Bửu was born on 23 July 1910, in Hoành Sơn Village, Khánh Sơn Commune, Nam Đàn District, Nghệ An Province. His home district of Nam Đàn contains Sen Village - the birthplace of Hồ Chí Minh - as well as Đan Nhiễm Village, home of patriot Phan Bội Châu. The area also includes natural and human-made landmarks such as the Lam River, Đụn Mountain, Mai Hắc Đế[1] Temple, and Đại Huệ Mountain, which is surrounded by clear brooks and farms that grow areca, jackfruit, orange, and persimmon trees.

Tạ Quang Bửu's paternal grandfather, Tạ Quang Oánh, passed the inter-provincial examination but returned to plow the village fields. In the records of the

1. A Vietnamese king who struggled against the Chinese Tang invaders in the early eighth century.

Tạ Quang family of Hoành Sơn, two short sentences have been preserved:

"Phụ giáo tử đăng khoa,
Cử nhân tại quán."

These sentences mean, "Fathers teach, sons sit for examinations." For each of the eleven generations of this family, there was a laureate. After passing the examinations, however, the honorees continued to live in their own villages and decided against becoming mandarins.

Tạ Quang Bửu's father, Tạ Quang Diễm, was famous for his studies from a young age. Intelligent and clever, he left behind many poems and witty songs that have been passed down to the present day. However, he also had a mischievous nature and liked to compose verses that irritated old-fashioned and narrow-minded scholars and mandarins. Therefore, Tạ Quang Diễm was failed at the examinations until he eventually passed the baccalaureate in 1910, when he was twenty-six years old. Since he knew many people in the movement of Đông Kinh Nghĩa Thục, an organization of reform-minded patriotic Confucian scholars, he was not selected to become a district mandarin in a central or well-off area, but was assigned to serve as an education officer in Quý Hương (Thanh Hóa Province) – a village with an unhealthy climate and poor people. Later, he was transferred to Tam Kỳ Prefecture (Quảng Nam Province) to work as a teacher. Being famous as a learned man, he was invited to serve on the examination board of Trường Thừa (Huế).

After World War I (1914-1918), France abolished the old examination system and Tạ Quang Diễm taught Han (classical Chinese) and Vietnamese literature in primary schools for a very small salary. His wife, Nguyễn Thị Đào, taught home economics and occasionally sent poems about the local scenery to the newspapers *Tiếng Dân* (Voice of the People) and *Phụ nữ Thời đàm* (Women's Commentary on Current Events), under the pen-name Sầm Phố. Tạ Quang Diễm and Nguyễn Thị Đào had two sons, Tạ Quang Bửu and Tạ Quang Đệ (who later became the famous journalist Quang Đạm). Tạ Quang Diễm died in 1925 when he was only forty-one years old.

His son, Professor Tạ Quang Bửu later recounted, "My father passed away on the sixth day of the fourth lunar month. At that time, I was staying in Nguyễn Thúc Dinh[1]'s house so that I would be closer to the National School (École Primaire Supérieure) where I was studying. Suddenly, someone came in calling for me to return home because that morning my father had been vomiting blood and had collapsed in the road, though luckily someone was there to help him up. Father died, but I didn't understand anything at all. I didn't yet know how to love my father in a profound way. My father had contracted tuberculosis because of poverty. After he was gone, there was nothing left for my mother except several hundred piastres in debts! Women and children in the streets told me I had to wear mourning clothes and a white headband, and lie on the ground waiting until people carried my father's coffin past. Only then could I get up. After that, the coffin was taken down to the boat and passed under the Thanh Long bridge and then was taken up again and buried in Bến Ngự. From that point on, I knew we would have it hard. Some of my father's friends showed me a way to receive a scholarship. With my clothes and headband tidied up, I

1. Deputy Minister of Home Affairs under the Nguyễn Dynasty.

took an application to the French Resident Superior. Under the friendly guidance of Lê Thanh Cảnh, I went to a room where there was one young Western man, probably not yet twenty-five years old, sitting. I humbly approached him with the application in my hand. I couldn't see the Westerner doing anything except using his hand to pluck his moustache. Suddenly something hit my ear like a blow from heaven! It hurt so much that I only vaguely remember a Vietnamese person stepping into the room and saying to the Westerner in French, "Why did you hit him? He didn't remove his mourning headband, that's not the same as refusing to take off his hat!"

This incident marks his first memory of feeling resentment about being a citizen without a country. He was only 15 years old.

Spirited Life as a Student

In 1917, there was an examination held in Tam Kỳ Prefecture for seven-year-old students. It covered Han characters, mathematics, and Vietnamese literature. The examination papers were marked by Confucian scholar Dr. Đinh Văn Chấp. Tạ Quang Bửu sat for the test and received high marks. From then on, he became known as an excellent student. He studied in Quảng Nam for four years, then moved to study in Huế and finished primary school in 1922. He came in eleventh in the entrance examination for the National School. Tạ Quang Bửu recalled, "I was placed in the A first-year class, based on alphabetical order. The teachers who taught us were Frenchmen, most of whom had a brevet supérieur (equivalent to today's junior secondary school diploma). There were very few teachers with a senior secondary school diploma or bachelor's degree. They taught mechanically, trying only to make students memorize classical principles. We took the classes that counted for the highest credit at term exams, final exams, and graduation exams. We also studied hard for any teacher who knew how to respect the students, who thought students were people too, and who did not box our ears or kick us in the rear. I generally liked history and geography because these subjects were very useful for learning French well. I did not like mathematics because M. Dubois often hit students, and M. Bruel often spoke nonsense. I was crazy for physics and chemistry. M. Surrugue introduced me to the torch battery, a simple device for generating electricity. The coins of Emperor Khải Định were made out of tin or copper. Silver coins at that time were made from real silver. Copper wire was also easy to find. A 2.5-volt light bulb was not hard to get, either. So there was a battery and electricity, and a light bulb. I liked electricity because it could "travel" far. Those were the romantic days of my youth. Out of all the teachers who taught us, there were a few Vietnamese, including M. Hoàng Gia Đức, who had passed the baccalaureate in France and read French very well, even better than the Western teachers! He chose wonderful beautiful French passages to read to us. We listened without breathing, like we were entranced. I can remember to this day hearing the teacher reading lyrical poems by Hugo or Lamartine, or transparent prose passages by Flaubert, Maupassant, or Daudet.

Tạ Quang Bửu was the youngest student in the class and was often teasingly called "the Lilliputian" by his classmates. He loved literature, but it became clear that he had an inclination towards natural sciences.

"We liked the wireless (radio)," he recounted further. "I found a telephone receiver. So I planted two bamboo poles in the garden and then looked for two porcelain knobs and fastened them together with copper wire to make an antenna. Then I connected the antenna to the telephone, held the receiver up to my ear, and pretended to listen – but didn't hear anything! Could this be called a kind of 'thought experiment'? Because of this curiosity and imagination, when I later studied quantum mechanics and encountered the 'thought experiment', it was easier for me to visualize it and get used to it."

In summer 1926, he came in second in his final examinations. He went to Hà Nội to continue his studies. After three years at the Lycée du Protectorat (Bưởi School), he passed the local baccalaureate with the highest marks in Hà Nội. Immediately afterwards, he took the Western baccalaureate exam and again placed first. The student with the second-highest score that year was a French student named Hiéroltz, who was admitted to the École Polytechnique in Paris, a school with a very difficult entrance examination. This demonstrates the quality of the baccalaureates trained in Indochina.

Encountering European Culture

He received a scholarship from the Hội Như Tây du học (Western Study Abroad Association) to go to Paris and enter a special mathematics class at the Lycée Louis le Grand. This school was an opportunity for the preparation for students who hoped to be selected for well-known schools such as the École Normale Supérieure, École Polytechnique, or universities of civil engineering, electricity and mining.

Even prominent scholars felt honored to be alumni of these schools. In 1930, he entered the bachelor's program in science at the Sorbonne. He attended lectures at the Institut Henri Poincaré, where there were two well-known lecture-halls: Hermite Hall, which was reserved for undergraduate students, and Darboux Hall for graduate students. The Vietnamese student was so eager to learn that he not only went to Hermite Hall but also attended seminars held in Darboux Hall. There, he attended lectures given by many famous professors like De Broglie, Fréchet, and Élie Cartan. At the time, many young French mathematicians (among whom were Tạ Quang Bửu's classmates Henri Cartan, Jean Dieudonné, and Godement) secretly joined the N. Bourbaki group, which aimed to write a thorough, unified account of all mathematics. In 1961, Professor Tạ Quang Bửu wrote the book *Về cấu trúc của Bourbaki* (On Bourbaki's Structures) published by Khoa học (Science) Publishing House, printed in Hà Nội.

Several years later, he left Paris for Bordeaux in order to further study mechanics further with Professor Trousset. He carefully read Rauth's *Mechanics* and completed most of the exercises therein. Knowledge from his youth helped him to later become one of the founders of the Việt Nam Association of Mechanics. He wrote many reports for seminars of the association (his last report was *Biological Mechanics*, which was planned for presentation in September 1988). Next, he applied and was accepted to Oxford University, a celebrated center of science in England. Receiving a scholarship from Oxford, he left France for England. There, he had the opportunity to improve his English skills and further his study of quantum physics through seminars.

Five years later, he had a chance to return to England. This time, he entered the Scoutmaster Training School near London and took examinations for training Boy Scouts in skills such as camping, tying knots, and cooking. Upon completion he received a Scoutmaster certificate for all of Indochina. He had participated in the Scout Movement since 1937 because he felt that the movement was "clean" and had a somewhat patriotic character. It contained activities aimed at building up character traits in youth and young intellectuals such as resourcefulness, self-reliance, and charity. Disliking the French, he looked for ways to gradually remove French influence from the Scout Movement and to quietly counter the Ducoroy Movement. Many scouts later secretly joined the Việt Minh. Members of the General Scout Council of Indochina, such as Tạ Quang Bửu and Hoàng Đạo Thúy, became leading officers in the Revolutionary Army of Việt Nam.

In 1946, he returned to Paris as a member of the Vietnamese Government's Delegation and attended the Fontainebleau Conference. In July of that year, with the approval of the head of the delegation, Phạm Văn Đồng, Professor Tạ Quang Bửu and Dr. Bửu Hội went to Geneva to attend the 200th anniversary celebration of the Swiss Association of Natural Sciences. This was the largest scientific conference organized in Europe after World War II.

"There," the professor recounted, "we met almost all the great Western European scholars, some of whom had experienced great difficulties when their countries were occupied by Hitler's forces. At that time, the Democratic Republic of Việt Nam was not yet two years old, but there were already two scientific officials attending that conference: Dr. Bửu Hội, who was working at the French National Center for Scientific Research, and me, a professor of physics at the Hà Nội Science University." The anniversary celebration took place at the Zurich Polytechnic University, where Albert Einstein was trained and where prominent scholars and Nobel Prize winners like Schrodinger, Pauli, and Karrer taught and conducted research.

Let us listen to Professor Tạ Quang Bửu recount how Erwin Schrodinger was received, "He is an Austrian physicist who was awarded the 1933 Nobel Prize in wave mechanics. He had taught at the Zurich Polytechnic University for a long time before he went to Ireland to teach and conduct research. Returning to Zurich this time, he was welcomed by everybody as the 'eldest brother.' He gave an intimate but profound talk in his easy-to-listen-to and beautiful Austrian voice. Before I came there, I had read many of his books on physics, especially *Was ist Leben*? (What is Life?), which was written in German and published in 1944. He gave me many profound suggestions which, even now, still have a beneficial influence."

The professor also recalled another anecdote, "What I now remember most is a talk about bee language given by Austrian scholar Karl von Frisch, who was then nearly sixty years old. Most of the audience members were not entomologists and I was not very good at German. But everybody listened with interest and felt that they were witnessing the birth of something great. Later, I knew that what was born that day was behavioral science. However, Karl von Frisch did not receive the Nobel Prize until 1973, when he was eighty-eight years old." He remembered the witty words of scholar Paul Karrer who won the Nobel

Prize for discovering and synthesizing vitamin B2. He also remembered his motorboat tour on the waveless Lake Zurich, the verdant hills, bustling industrial zones in Baden and Winterthur, and the friendships with those who he worked with in the scientific field.

He said that his visit to Zurich motivated him to write his four books that were printed by the Defense Printing House in the forests of Việt Bắc in 1948. These books were *Nguyên tử, hạt nhân và vũ trụ tuyến* (Atoms, Nuclei and Cosmic Rays), *Sống* (Life), *Vật lý cương yếu* (Fundamentals of Physics), and *Thống kê thường thức* (Basic Statistics). These books had a deep influence on young intellectuals at that the time.

Not long before his death, he still reiterated, "In my opinion, young scientists (in 1946, I was thirty-six years old and Bửu Hội was just over thirty) should be offered a chance to attend large international conferences. Of course, they must have enough knowledge of English and Russian to understand." As many people know, by the time he was twenty-four years old, the young scholar from Nghệ An Province had already mastered French and English, could use German, and could read ancient Greek and Latin. Later, he added Russian, Chinese, and Polish to his repertoire.

Youthful Strength Despite Age

Because of the great advances of medical science, the today's elderly are able to live youthful and happy lives, which are, in turn, beneficial to society. I felt this fact particularly strongly on the nights in late 1986 when I was reading a type-script of Tạ Quang Bửu's book *Hạt cơ bản*[1] (Elementary Particles). This is a high-level book of popular science, easy to understand by anyone who has a university-level understanding of physics. The author guides the reader from the philosophical thoughts about atoms that Democritus had in ancient Greece up to the ideas of P.A. Dirac (1933 Nobel Prize winner) on the principles of quantum mechanics. By explaining three elementary particles (discovered in 1932, 1960, and 1980), he summarizes the physics discoveries of recent decades, from quantum electrodynamics to quantum chromodynamics. By reading the book, we can learn about unitarian symmetry, particles like the charm quark, beauty quark, and strange quark, and Weinberg-Salam's endeavor to unify the weak interaction and electromagnetic interaction.

He concludes, "Now we can see that the electromagnetic and weak forces have already been unified into one theory, namely the Weinberg-Salam theory. But, in reality, we do not yet have a complete theory about the strong and gravitational forces. If the dream of unifying everything into one theory can be accomplished, then we will only need to write one equation and quantify it accordingly, and the whole system of the cosmos – from micro to macro – will all be described by a single theory. People who are optimistic expect this work to be accomplished by 1990."[2]

He retired in 1979. In spite of this, he continued to observe the latest trends in physics, reading difficult-to-understand articles by A. Salam and S. L. Glashow, and explaining the discoveries of S.C.C. Ting (Ding Zhaozhong, a Chinese-American who won the 1976 Nobel Prize).

1. Giáo dục (Education) Publishing House, 1987.
2. As of 2010, this has still not yet been accomplished (author).

Up to the last minutes of his life, he still kept his intellectual youthfulness and emotional passion. His last book, which was printed after his death, is also reserved for young people – those who will continue to explore the micro-world and macro-world and who will discover the wonderful harmony and simplicity of an immortal nature.

Many Fields of Knowledge

Some people say that Tạ Quang Bửu is a modern-day Lê Quý Đôn[1]. This is a well-founded remark. When he studied in France, he was one of four people to pass a mathematics exam out of over 100 candidates. His learning method, which was different from the intensive studying methods practiced by many Vietnamese of his time, was to study in order to know rather than studying in order to perform well on tests. He liked painting and sports. He practiced running (from 100 meters to 10,000 meters), high jumping according to the straddle technique (a method in which the athlete's stomach faces the bar), long jumping in the scissors style, and swimming the crawl. He took part in a swimming contest across the Seine River, in which French champion Jean Taris also participated. He received a swimming certificate from the United Kingdom. He rose to the finals of the Parisian students' table tennis competition, using the method of Hungarian champion Barna. He also played soccer and practiced boxing for self-defence.

Tạ Quang Bửu also had a thorough knowledge of architecture and was a music connoisseur. He often sang "Ode to Joy" in German, whose words (by Friedrich von Schiller) had been set to the music of Beethoven's "Ninth

1. The first encyclopedist of Việt Nam (eighteenth century).

Symphony." Professor Hoàng Như Mai said, "During the Arts and Literature Conference held in Việt Bắc in 1949, there were many different opinions evaluating this or that facet of Nguyễn Xuân Khoát's music. Only the opinion of Tạ Quang Bửu came through clearly and decisively, 'This is just music.'" Regarding his architectural knowledge, Associate Professor Đặng Thái Hoàng said that when he completed his book *Lịch sử kiến trúc thế giới* (History of World Architecture), he asked Tạ Quang Bửu's opinion. He offered specific and profound suggestions, and Đặng Thái Hoàng had to make some corrections.

When he returned to Việt Nam, he did not accept the offer to become an official, but instead took a post teaching math and English at a private school, Providence School in Huế. He realized how hard it was to deepen his understanding of Vietnamese and Oriental cultures without knowing Han (classical Chinese). Thus he immersed himself in learning the most difficult system of writing in the world. He also learned from Phan Bội Châu in Bến Ngự. His study was immense. He began to read *Analects, Great Learning, Doctrine of the Mean, Mencius, Dao De Jing, Nan Hua Jing, Shiji, Romance of the Three Kingdoms*, and some poems in the original language. He studied French culture and Greek philosophy but also taught himself Chinese in order to thoroughly understand the literature, history, and philosophy of classical and medieval China.

Professor Tạ Quang Bửu was posthumously awarded the Hồ Chí Minh Prize for his collection of works to introduce science and technology (after 1945). At the 1948 National Culture Conference held in Việt Bắc, Professor Nguyễn Xiển, a veteran math teacher, said, "During this time of resistance, Tạ Quang Bửu is the scientist who writes the most, so he may have most influence on contemporary generations." He continued, "With such pioneers as Tạ Quang Bửu and Lê Văn Thiêm, Việt Nam will certainly have hundreds of mathematicians who are as talented as foreign mathematicians."

Professor Lê Văn Thiêm later recounted, "In 1951, after coming back from Switzerland to the resistance base in South Việt Nam, I walked up Trường Sơn Trail to Việt Bắc. One day I visited Tạ Quang Bửu in a thatched hut that served as his office in the forests of Tuyên Quang. I was pleasantly astonished to see that, though he was immersed in his work, he still found time to read well-known books and articles on mathematics in German, English, and French." According to Professor Lê Văn Thiêm, his self-study ability was legendary.

American linguist and mathematician Noam Chomsky, who was regarded by *Newsweek* as one of the most prominent mathematicians in the twentieth century, met Tạ Quang Bửu on his visit to Việt Nam. When he returned to the United States, Noam Chomsky wrote, "Mr. Tạ Quang Bửu is a person of formidable intelligence."

Professor Tạ Quang Bửu was accustomed to reading books in any place and at any time. Although he read very quickly, he was able to remember and retain the information on what he had read. Once, he was so absorbed in reading while riding a horse that he fell onto a stream!

He also mastered many foreign languages. Mr. Nguyễn Xuân Huy said that before the Điện Biên Phủ Operation, soldiers received Soviet anti-aircraft missiles along with instructions for their use that were written in Russian. Mr.

Nguyễn Xuân Huy did not know whom to ask to translate the instructions, since at that time the Ministry of Defense only had Chinese interpreters. Hearing that after three months of self-study Tạ Quang Bửu could read Russian, he walked for a day and a night to meet him. "Tạ Quang Bửu looked at it and read the Russian without a pause. I respected him so much! Then he translated it immediately into French for me to take back for research and to guide the soldiers," he recalled.

In 1963, Polish mathematician Mikusinski sent Professor Tạ Quang Bửu his new research results. Tạ Quang Bửu read the Polish version and then gave a lecture on Mikusinski Operator to the math lecturers of Hà Nội universities.

Tạ Quang Bửu spoke excellent English. One time, he talked about Protestantism in England so beautifully that people thought he was a pastor, even though - according to him - he was only a "believer in patriotism." In the first days after the August 1945 Revolution, when he was serving as an advisor to the Ministry of Foreign Affairs (at that time, Hồ Chí Minh himself served as Minister of Foreign Affairs), he helped Uncle Hồ write diplomatic dispatches to Stalin, Truman, and Attlee and receive American and British diplomats. In his 500-page book *Why Việt Nam?*, Archimedes L.A. Patti, an American intelligence colonel who observed people and events in Hà Nội in 1945, wrote the following, "A guest was waiting for me at the villa. It was Mr. Tạ Quang Bửu, a Vietnamese who seemed intelligent and was about thirty years of age. I thought I had met him somewhere but was not sure. Bernique and I shook hands with him, and he introduced himself, saying that he was sent by the Ministry of Interior. His perfect English, with a characteristic Oxford accent, astonished me."

On March 6, 1948, Tạ Quang Bửu's book *Sống* (Life) was published in the forests of Việt Bắc. The book's title was short, but the thoughts it contained were far-reaching. Over fifty years later, when I read the thin book printed on *dó* paper, I was surprised to realize that in those days, the Vietnamese erudite could have used the latest discoveries of quantum physics to explain life and to represent gene structure, hereditariness and variability, dominant and recessive traits, mutagens like Roentgen, and cosmic rays. Planck's, Dirac's, Heisenberg's, and Schrodinger's ideas about the micro-world were also included in the book.

The last section of the book has a philosophical flavor. The professor writes, "Germs are alive. Flowers and pigs are also alive. But does the flower know it is alive? Life is one issue; self-awareness is another. According to Schrodinger, if I and the cosmos are one, then I am a small part of the cosmos. So then I do not die, because the cosmos does not die!... Of course there are people who will object: Whether my atoms live or not is irrelevant! What I want to know is whether I will live forever or not." He responds, "The critical point is not 'What is life?' The most critical point is, 'What should we do while we are alive?'" He must have thought very hard about this topic.

Professor Tạ Quang Bửu ended his work on the evening of August 14, 1986, and suffered a stroke. One week later, at noon on August 21, he departed this life, leaving many unfinished writings and unpublished manuscripts. *Human Strategy* was an unfinished manuscript that was to be presented at the Sixth Congress of the Communist Party of Việt Nam.

Recently, Hà Nội has a new road, Tạ Quang Bửu Street. It links Bạch Mai and Đại Cồ Việt through the Polytechnic University.

TÔN THẤT TÙNG – THE FATHER OF LIVER SURGERY

Recognition From Around the World

Tôn Thất Tùng passed away on May 7, 1982, the twenty-eighth anniversary of the victory at Điện Biên Phủ in which he had taken an active part.

On a wide bed in the middle of the room, the professor lay with his arms outstretched. A blue blanket covered him. His head rested on a soft, white pillow, his silver hair falling down around his temples. The dark green walls, lacquer paintings, and wooden shelves loaded with the books and scientific journals he read daily, and the beret he used to wear, were all unchanged.

Sitting at his bedside, Dr. Tôn Thất Bách, Professor Tôn Thất Tùng's son caressed his father's forehead and cheeks. "Yesterday evening, my father was still joking and laughing with us. He worked late until 1:00 a.m. In the morning, he felt tired and told me to wake him up at half past ten. But at 11:00 a.m., he was still not awake..."

A woman from the countryside, holding her young son with her left hand and a bundle of incense sticks in her right, entered the room. Seeing the professor's face looking pale, she burst into tears, "Oh, Uncle Tùng! You are the one who's saved me and my son!"

The sad news spread rapidly. Thousands of relatives, friends, students and patients came to the house near the Hà Nội Medical University to pay respects to this beloved doctor for the last time. They pressed together in the garden, on the steps, or outside the fence waiting for their turn.

Dr. Hồ Đắc Di said in a subdued voice, "I've lived in the same house as Tùng for more than half a century, so I've known him pretty well. He is extremely kind. Out of all the surgeons in the world, very few have ever received the Lannelongue Medal - even fewer than the number of Nobel Prize-winning physicists or Field Medal-winning mathematicians. One French professor even told me, 'Tôn Thất Tùng is a luxury for Việt Nam!'"

Dr. Đỗ Xuân Hợp, the former rector of the Military Medical Institute, recalled, "I was responsible for training residents at the Institute of Anatomy beginning in 1932 and I knew Tùng since he was a fourth-year student. He had passion for science and spared no effort for observation and practice, out of which he built up new theories. Detecting that worms had infected a liver, he began to dissect (the liver) and then patiently continued the dissection of several hundred more livers, delineating for the first time the whole network of blood veins and gall ducts within the liver. He then proposed a new method for liver surgery. Later, he conducted research into treatment of liver cancer and the toxic effects of dioxin."

At his funeral, friends and students came to Hà Nội from all corners of the country to bid farewell to the professor.

Thousands of patients he had cured and hundreds of young doctors and medical students wept before his coffin. Foreign embassies in Hà Nội, the TASS news agency, Denpa News (Japan), UNICEF, and American doctor Levinson all sent funeral wreaths.

In their telegram of condolences, the Soviet Minister of Health and the President of the Medical Academy of the Soviet Union wrote, "Medical circles in our country have known Tôn Thất Tùng as an outstanding educator who has made important contributions to the medical development of the Socialist Republic of Việt Namand and to medical cooperation between our countries."

And a French professor, Dr. Lederer added, "I learned of the death of my friend Professor Tôn Thất Tùng in *Le Monde*. He was an eminent surgeon and a great patriot whom I met frequently. His passing is a great loss for humanity and for your country in particular."

In his article, "The Death of Tôn Thất Tùng - A Great Surgeon Passes Away," Dr. Jean-Michel Krivine, a surgeon at Eaubonne Hospital in France, wrote, "Tôn Thất Tùng was a man of dialogue, non-partisanship, proud of his double French-Vietnamese culture.... Thanks to him, medical exchanges between our countries have intensified. 'The French doctor Meyer-May gave me good surgical and clinical training as a generalist,' he liked to say. 'For the rest, I taught myself.' Such a life is truly exemplary. May surgery in Vietnam continue to be inspired by the enthusiasm, the seriousness and the open-mindedness of its founding father."

From the United States, Dr. Judy Ladinsky, Dr. Glenn Gordon, and Dr. Ngô Vĩnh Long expressed "our deepest sympathy on the passing of a great man and an inspiration to all of us." From Italy, the young surgeon Guido de Sena at the Naples Hospital sent a short telegram to Hà Nội, "I've lost my second father." And from Cambodia, Dr. Sam Sophean wrote, "We are deeply moved to hear the news of the passing of our beloved professor."

Scientists over the world and overseas Vietnamese associations organized memorial services. At a ceremony in Paris, Bùi Mộng Hùng, chair of the Vietnamese Medical Association in France, expressed his "condolences on the occasion of the death of Prof. Tôn Thất Tùng, an intelligent, creative and devoted teacher and colleague, who was always friendly towards overseas Vietnamese physicians."

And finally, in his preface to a book by Tôn Thất Tùng entitled *My Path to Science*, General Võ Nguyên Giáp wrote, "Professor Tôn Thất Tùng is a scholar of great renown. And as a medical advisor to the High Military Command during the anti-French resistance war, he was an outstanding warrior on the scientific and technological front for our country."

Memories of the War of Resistance

In addition to his scientific works, Tôn Thất Tùng left an autobiography, a diary of which a number of pages were written at the Điện Biên Phủ battle field, and poems in French that few people knew about, some of them translations from Vietnamese.

During the anti-French resistance war, the Hà Nội Medical University was evacuated to Chiêm Hóa (Tuyên Quang Province). Professor Hồ Đắc Di took over the responsibilities of rector. Professor Tôn Thất Tùng was

Director of the Field Hospital. In 1947, Professor Tùng was named Vice-Minister of Health by President Hồ Chí Minh. Every month, he had to bicycle some 200 kilometers through the forest and past Bình Ca Wharf and Khế Pass to attend the meetings of the Government Council, presided over by Uncle Hồ.

In 1949, Tôn Thất Tùng collaborated with Professor Đặng Văn Ngữ to produce penicillin in Việt Bắc. This was a miracle. In the history of guerrilla wars throughout the world, no one had ever succeeded in achieving this task with such rudimentary tools. As Professor Tôn Thất Tùng later recalled, "Appreciating this success, Uncle Hồ told me once at a Government Council meeting, 'It's up to you to choose the medal that you want!' Since comrade Ngữ had received the Order of Resistance, Third Class, I requested the same. Several days later, Uncle Hồ invited the members of the Council to dinner in order to present the awards. He said, 'Tùng is a *ci-devant* (a term for aristocrats used during the French Revolution), but today he received a medal from our government. This will encourage him to make more effort.' This is the most beautiful memory of my life."

On March 23, 1954, Professor Tôn Thất Tùng received a letter from the High Military Command directing him and Dr. Vũ Đình Tụng to go to Điện Biên Phủ.

"That morning, I received a letter," he recalled. "In fact it was a dirty piece of paper put in an inside-out envelope with a few words about our immediate departure for Điện Biên Phủ with a familiar signature beneath. In the afternoon, when the enemy airplanes "went to sleep," I got in a basket boat with my surgical team. We traveled down the Lô River, through countless forests and rapids, to reach our destination. We knew that the Điện Biên Phủ campaign was entering its second phase..."

In Tuyên Quang, the team found a truck waiting to take them to Chợ Hiên. Along the road, thousands of conscripted laborers were carrying supplies to the front. At night, they pushed their heavily loaded bicycles through the cold rain, crossing streams and passes.

At the end of March 1954, after surmounting the cloud covered Pha Đin Pass, Professor Tôn Thất Tùng arrived in Điện Biên Phủ. He lived and worked in a shack on the bank of a roaring stream. The battle of Mường Thanh was raging and the wounded were becoming more and more numerous.

The professor recorded in his diary, "April 5, 1954 (Surgical Station): This morning I visited the severely wounded. It rained all night. Thinking of the wounded in the trenches brought tears to my eyes and breaks my heart. I never had so much compassion for soldiers as today. Fortunately for the brothers at the front, in the morning, it stopped raining.

April 9, 1954 (Surgical Station 1): Last evening it rained. During the storm, thunder mixed with the sound of cannons. A falling tree killed two assistant doctors and a student.

10:00: We've operated on all the wounded in turn, despite incessant shelling.

11:30: The rain and winds have tired us out. I'm beginning to feel unwell. Yellow flies are biting our legs. This reminded me of an expression about the three curses of the Northwest, 'yellow flies, dog fleas, and the wind from Than Uyên.'

19:00: Exhausted, I couldn't finish a bowl of rice. Brain operations wore me out since I had to keep on injecting anesthetic and cutting skull tendons. My fingers and muscles ache. The number of wounded has risen to 700, but we only have six assistant doctors and twenty nurses. The whole team is worn out. We couldn't even eat..."

It rained all week, day and night. The soldiers waded through muddy trenches. One night, Professor Tôn Thất Tùng got a call at 12:45. It was an emergency. A young soldier, about twenty years old, said with a central Vietnamese accent, "Doctor, please don't hurt me any more!" His abdomen had been punctured, and his intestines were rotting. The professor gave him pre-operation injections, and then wrote in his diary, "If it weren't for this war of resistance, I would never have realized how brave our people are."

The following is another extract from the diary, "May 7, 1954: In the afternoon of the day before yesterday, I heard that the High Commander was coming to visit medical Stations 2 and 3 and one fighting unit. We were waiting for the car in a burnt-out village when a comrade passed us on a bicycle shouting 'Điện Biên Phủ is liberated!' It may be true, I thought, because we hadn't heard any shooting since noon. Then around six-thirty in the evening, Comrade Ch. arrived running. He said, 'Tùng, we've captured Điện Biên Phủ!' I hugged and kissed him and old Tụng. I yelled 'Bravo! Bravo!' like a madman. Then we left immediately to tell the news to our team.

Stepping in the sub-zone where the team was stationed, I asked a man nearby, 'Have you heard any news?'

'What news?'

'De Castries has surrendered!'

He jumped for joy, screaming, 'Is it true? Bravo! Bravo!...'

Some female conscripted laborers were yelling too, 'Bravo! Bravo!...'

A Surgeon With a Love for Poetry

Together with Phan Nhuận, Kỳ Anh, Jacques Gaucheron, and Georges Boudarel, Tôn Thất Tùng translated *Từ ấy* (Since Then), a collection of poems by Tố Hữu, into French (published by the Foreign Languages Publishing House, present-day Thế Giới Publishers, Hà Nội). He loved classic poets Nguyễn Trãi and Nguyễn Du, particularly those written in *Nôm* ideograms. He regretted never having learned to write poetry in his native language. When he was studying at the Lycée du Protectorat (Bưởi School) and at the Hà Nội Faculty of Medicine, he became accustomed to composing poems in the French language. He was heavily influenced by poet, mathematician, and scholar Paul Valéry and kept in a correspondence with him for many years.

As a man of science, he was also a humanitarian, as the following story demonstrates. On November 4, 1981, the Vietnamese-German Hospital in Hà Nội took in a 7-year-old boy whose arms and one leg had been crushed by the wheels of a tram. The doctor on duty wanted to amputate, but Professor Tôn Thất Tùng said, "With our equipment, we can handle all kinds of operations. But we have to think about the patient's future so that the rest of his life will be less difficult. With two amputated arms, he would be 100% incapacitated. If that's inevitable, we should look for every means to create favorable conditions for him."

Shortly thereafter, he contacted the charitable organization CARTINAS Germany to request a pair of artificial electronic arms for the poor boy. CARTINAS agreed.

Lasting Contributions to Medicine

Professor Tôn Thất Tùng was considered in international medical circles of medicine as a prominent surgeon. He was also a member of many surgical academies and associations. The Lannelongue Medal was given to him because he was first and foremost an innovator.

According to the French journal *Lyon chirurgical*, the Hà Nội Medical University should take pride in two innovations of Tôn Thất Tùng: the discovery of blood vessels in the liver and the codification of the technique of liver surgery.

The process of developing this technique, known as the Tôn Thất Tùng Method, lasted thirty-nine years (from 1935 to 1974). In the winter of 1935, he was a 23-year-old medical student. He used to go to the Institute of Anatomy, under the direction of Professor Huard, to dissect corpses. One day, he found a liver that had worms in all of its blood vessels and bile ducts. With nimble fingers and a scalpel, he managed to dissect all the ducts and vessels. Up to that point, only two methods of analysis were known: X-rays and acid corrosion. The drawback of X-rays was that all the vessels in the liver were rendered on a plane surface, which led to confusion. The second method was no more effective. Thus the vessels of the liver had never been accurately delineated. Dissection using a scalpel was an initiative of Tôn Thất Tùng. It only took him fifteen minutes to discover all the bile ducts and blood vessels and to describe them accurately.

Over the next four years (1935-1939), Tôn Thất Tùng embarked on a monotonous and gruesome task: dissecting the livers of more than 200 cadavers. He drew diagrams to compare them and find their common traits. On this basis, he completed his doctoral thesis, *Classification of Blood Vessels in the Liver* (La vascularisation veineuse de foie), which was awarded a Silver Medal by the University of Paris, of which the Hà Nội Medical University was then a branch. About this feat, he later recalled, "When I was preparing the thesis, Professor P. Huard wanted to point me towards anthropological morphology. He had prepared his own hypotheses and wanted me to do research and prove them correct. If my research reached a different conclusion, my future would be threatened. I was discouraged to find out that the reality of my life was quite different from that described in books. I understood how difficult it was to be objective in science. Quietly, I quit working on the direction that Professor Huard had suggested, without letting him know. Since then, I concentrated on studying the blood vessels in the liver."

In 1939, encouraged by his findings, the young surgeon of twenty-seven proposed a daring innovation: blinding together the blood vessels in the liver and then performing surgery. This method, later called the "standard technique of live surgery," was first employed by Tôn Thất Tùng. Before him, German, British, and Russian surgeons had conducted a total of eighty-seven hepatectomies, but these had not been standardized due to the lack of an exact description of the liver vessels. In these cases, the surgeons bound all the vessels they saw. However, if they accidentally missed one, the patient would die of hemorrhage or gangrene.

The first report on standard liver surgery conducted by Tôn Thất Tùng and Meyer-May was presented to the Paris Surgical Academy. However, it was rejected. According to Dr. Krivine, the idea was so new that the scientific establishment could not accept it. Disappointed, Tôn Thất Tùng did not touch any liver with his forceps for the next twenty years.

Only after the liberation of Hà Nội did he learn that in 1952, French Professor Lortat-Jacob had successfully executed a hepatectomy by tying up the stems of blood vessels *outside* the liver before surgery. Excited by this news, he resumed the work that he had abandoned in 1939.

On January 7, 1961, he removed the right lobe of a patient afflicted with liver cancer in only six minutes. The Lortat-Jacob method would have taken three or four hours. The difference was that he found the vessels and ducts *inside* the liver, through the destroyed liver tissue.

Soon thereafter, he published a short article in the London-based journal, *The Lancet*. "A New Technique for Operating on the Liver" astounded the world's medical circles. Within a month, over a hundred Western specialists from the United States to Australia sent letters to Hà Nội asking for reference documents. Some welcomed the new technique with reservations, while others categorically rejected it. Even a number of Vietnamese health officials were led to believe that his work was "anti-scientific." The new technique seemed to have been strangulated.

As he later recalled, "A researcher must brave, never discouraged or afraid of contrary opinions, persevering in work, and daring to argue. One must not have an inferiority complex, retreat in the face of rejection from Westerners, or even abandon the effort if some considers it 'anti-scientific.'"

Now he felt more self-confident. If in 1939 he was discouraged by the unjust criticism of the Paris Surgical Academy, by 1962 he was ready to throw himself into an unequal argument with great names in the world's medical circles. In the end, he was the winner. Once they understood the new technique, his most severe critics changed their tune, extolling his as the "father" and "master" of standard liver surgery.

His books *Chirurgie d'exérèse du foie* (Hepatectomy) was first published by *Masson* in France and then by *Medicina* in the Soviet Union. It was sold out in the USSR, Germany, France, Britain, the Netherlands, Norway, Sweden, Switzerland, Italy, and the United States. His work was also printed in *Obstetrics and Surgery Reader's Digest* in America. The "Tôn Thất Tùng Method" was presented in the *French Encyclopédie médico-chirurgicale.*

In 1979, *Masson* Publishers in Paris printed a new work by Tôn Thất Tùng entitled *Major and Minor Excisions of the Liver*, in which the professor summed up 715 cases of liver surgery he had performed. This figure far surpasses any figure that other surgeons in the world have recorded so far (after Tôn Thất Tùng, one Singaporean surgeon has executed over 100 cases).

Dr. Jean-Michel Krivine states, "In this book, Professor Tôn Thất Tùng's method is definitely codified. One can truly say that he has 'democratized' liver surgery, allowing any general surgeon who takes the trouble to learn to no longer be disarmed by a case of liver failure."

In 1977, Professor Tôn Thất Tùng was awarded the Lannelongue Medal, an international surgical medal from the Paris Surgical Academy. This prize had been created in 1911 and is awarded only once every five years to the world's most celebrated surgeon over that period. He was the twelfth person to be awarded the medal.

In additional fields, notably research on dioxin, he also registered new discoveries. He stated that gallstones in Việt Nam were often found not in the gall bladder as in Europe, but in the liver. Pancreatic failure in Việt Nam, he found, was usually not because of stones but because of worms invading the gall bladder. He also discovered a new disease and named it *tropical hemobilia* in order to distinguish it from similar diseases in Europe and America.

One week after his death, the Parisian newspaper *Témoinage* (Witness) wrote in its May 14, 1982 issue, "During American bombing, Tôn Thất Tùng found the time to study the tragic effects of defoliants, notably dioxin, on the population. He became a world-renowned specialist on dioxin; the Italians called on him after the explosion of the factory in Seveso.... Tôn Thất Tùng was not only a patriotic surgeon. He had an encyclopedic mind, and his conversation in impeccable French always charmed his interlocutors. Deeply good and warm, he had a commanding personality. His colleagues both loved and feared him."

In order to devote himself completely to his profession, Professor Tôn Thất Tùng resigned from the post of Vice-Minister of Health that he had held since 1947. In 1958, he carried out the first successful heart operation in Việt Nam. In 1964, ten years after the liberation of Hà Nội, he managed to restore a dilapidated hospital built under French domination into a modern facility with 350 beds and sixty specialists and assistants, of which a number could be classified as professors according to international standards. Many services of the hospital were on a par with those in other countries. In 1965, for the first time in Việt Nam, he successfully employed a cardiopulmonary machine in a heart operation.

To resolve difficult issues, he did not only consult with his Vietnamese colleagues. In 1970, Prime Minister Phạm Văn Đồng assigned him with the task of investigating the negative effects of the herbicides and finding remedies. Knowing that his knowledge of biochemistry was insufficient, he contacted Bửu Hội, a renowned biochemist who was working at the French National Scientific Research Center, as well as professors Pfeifer, Galston, and Wald (a Nobel Prize winner) at Harvard University and other biochemists from Britain, France, Italy, and Sweden.

When experimenting on mice suffering from liver cancer, he noticed that the level of angina in the liver was low. How would this be in a human liver? He requested copies of all documents that had been published on the subject between 1965 and 1973. Out of a million documents, the foreign archives found only seventy-five that mentioned enzymes in liver cancer patients and none at all that mentioned angina in liver cancer in humans. He knew that this was a new matter. Up until his death, he remained a researcher who was full of new ideas and could quickly grasp and apply the most advanced ideas to his profession.

After his death, foreign surgeons came to Hà Nội to learn his methods, among them the Italians Paletto, Salizzoni, and Giordani, who arrived in 1984. Dr. Salizzoni returned later to film the process of liver surgery carried out according to the Tôn Thất Tùng Method. Dr. Tôn Thất Bách (Professor Tôn Thất Tùng's son) and Dr. Nguyễn Văn Vân performed the demonstration. One accident that often occurs during liver operations is a cut by the scalpel on the lower vein. If the vein is not rapidly mended, the patient will die from loss of blood. Repairing the vein requires a surgeon with hands as nimble as those of a violinist. Book knowledge does not suffice. To everyone's astonishment, Dr. Tôn Thất Bách performed this task effortlessly.

During their sojourn in Hà Nội, the Italian doctors also completed a translation of Professor Tôn Thất Tùng's works, which appeared in Italy with updates from new applications of the professor's method. In autumn 1984, their film was shown at the International Liver Conference in the presence of two Vietnamese surgeons.

Professor Tôn Thất Tùng was the first scientist to denounce the spraying of herbicides in Việt Nam by America. The spraying increased the rate of liver cancer. In his book, *Herbicides in War: The Long-Term Ecological and Human Consequences* (1984), Professor Arthur H. Westing of the International Institute for Peace Research in Stockholm printed a black-bordered inscription on the title page, "This book is delicated to the soul of Professor Tôn Thất Tùng (May 10, 1912-May 7, 1982)."

Tôn Thất Tùng received the posthumous Hồ Chí Minh Order and the Hồ Chí Minh Award from the Vietnamese Government. His name has also been given to five streets in five cities in the country: Hà Nội, Huế, Đà Nẵng, Hồ Chí Minh City, and Long An.

Professor Hồ Đắc Di (right) and Professor Tôn Thất Tùng at the Medicine and Pharmacy University in the resistance base in the forest of Chiêm Hóa, Tuyên Quang Province, 1948.

President Hồ Chí Minh with heroes and model workers at the First Festival of Model Workers in Chiêm Hóa, Tuyên Quang Province, 1952. The third person from the left is Professor Trần Đại Nghĩa.

Professor Tạ Quang Bửu,
then Vietnamese Deputy Minister of National Defense,
signs the Genèva Agreement to restore peace in Indochina (1954).

Doctor Phạm Ngọc Thạch (in light-colored suit),
then Vietnamese Minister of Health, visits an antiaircraft artillery unit
in the anti-American national salvation resistance.

Professor Nguyễn Văn Huyên, a brilliant Vietnamotologist, with his wife, Mrs. Vi Kim Ngọc.

Top-ranking General Võ Nguyên Giáp
with Professor Nguyễn Xiển (right) and engineer Lê Tâm (middle).

NGUYỄN THÚC HÀO – A DEVOTED MATHEMATICIAN AND EDUCATOR

The majority of the founders of the new education system established in the independent Việt Nam after the August 1945 Revolution have already departed this life. In medicine, these people include Hồ Đắc Di, Tôn Thất Tùng, Đặng Văn Ngữ, Trần Hữu Tước, and Đỗ Xuân Hợp; in social sciences and humanities, Đặng Thai Mai, Đào Duy Anh, and Cao Xuân Huy; and in physics and mathematics, Tạ Quang Bửu, Lê Văn Thiêm, Ngụy Như Kon Tum, and Hoàng Xuân Hãn.

The surviving members of this generation are now all in their nineties. Prominent among them is mathematics teacher Nguyễn Thúc Hào, who has been granted the honorary title of "People's Teacher." He was born on August 6, 1912 into an educated family in Xuân Liễu Village, Nam Đàn District, in the province of Nghệ An, rich in patriotic and cultural traditions. His father was Doctoral Laureate Nguyễn Thúc Dinh. At the outset of the twentieth century, the colonial administration was solidly established. He thus belongs to the first generation of Vietnamese intellectuals with French university training.

He began teaching mathematics at the National School (later Lycée Khải Định) in Huế in 1935. He clung steadfast to the teaching profession after the August Revolution until his retirement in 1976. He was convinced that mathematics plays a fundamental role in the formation of scientific thinking among young people in a country molded by Confucian philosophy.

Professor Tạ Quang Bửu once recounted, "He and I come from the same district of Nam Đàn. In the 1924-1925 school year, I was in the third year at the Lycée Khải Định in Huế. His father let me live in his house so that I could be closer to my school. A good traditional scholar, he taught his sons Hào and Tùng (who became an eminent medical doctor) in a strict and orderly manner.... The next year, when I was in the fourth year, Hào and Võ Nguyên Giáp were accepted into the first-year class with the top scores in the entrance exams. Giáp at that time was a small but intelligent student."

Nguyễn Thúc Hào attended the school in Huế only for one year. Between 1926 and 1929, he went to the Lycée Albert Sarraut in Hà Nội. Then he went to France and completed his studies in Aix-en-Provence and in a pre-university course at Saint-Louis in Paris. However, pulmonary disease forced him to leave the cold in Paris to enroll in the University of Marseille located in the south of France. During the four years from 1931 to 1935, he obtained six certificates in general mathematics, analytic mathematics, general physics, theoretical mechanics, fluid mechanics, and astronomy. In addition, he completed a master-level thesis related to geometry and mechanics.

He returned to Huế at the age of twenty-three and began to teach at the Lycée Khải Định. At that time, there were six secondary schools in French Indochina (the others being the Lycée Albert Sarraut and Lycée du Protectorat in Hà Nội, the Yersin School in Đà Lạt, and the Chasseloup Laubat and Petrus Ký Lycée in Sài Gòn).

Between 1942 and 1944, some young French-educated scholars including Nguyễn Xiển, Hoàng Xuân Hãn, and Đặng Phúc Thông published the journal *Khoa học* (Science) in Hà Nội. The journal drew the interest of both domestic and overseas readers. Intellectuals in Huế such as Nguyễn Thúc Hào, Tạ Quang Bửu, and Phạm Đình Ái maintained fruitful collaborations with them. They also took an active part in the Association for the Dissemination of *Quốc ngữ*, the Romanized Vietnamese script, holding that illiteracy eradication was the cornerstone of the struggle against colonialism.

The former imperial capital of Huế, which had been sleeping for a century on the banks of the Hương (Perfume) River, awoke as the August 1945 Revolution broke out. In the exciting atmosphere those days, Nguyễn Thúc Hào worked with unimaginable enthusiasm. In addition to teaching math, he assumed the position of Director of Secondary Education of Central Việt Nam and became a member of the Council of Education Advisors for the Ministry of Education of Independent Việt Nam.

During a meeting of the council in Hà Nội, he had a chance to dine at the home of an old acquaintance of his. This colleague recalled the days they spent in Huế and told him how, after teaching at Thăng Long School in Hà Nội, he joined the resistance in China with Hồ Chí Minh until the victory of the revolution. That old friend was Võ Nguyên Giáp.

By February 1946, the young republic faced many economic, political, and social difficulties. At that time, the student movement was divided along political lines among students loyal to the Việt Minh and those apart of rival political parties such as the National Party *(Quốc dân đảng)* whose leaders returned to Việt Nam with Chiang Kai-shek's (Jiang Jieshi's) troops. The Ministry of Education decided to close temporarily higher educating establishments until the following autumn in order to allow students to participate in more urgent activities and to ensure better order and discipline in the higher education system.

Nguyễn Thúc Hào was called back to Hà Nội to take up the position of General Secretary and Acting Director of the Hà Nội Science University in place of Dr. Hoàng Thị Nga who had immigrated to France. He tried his best to reorganize classes, recruit students, design a new curriculum, and assure the cooperation of competent professor such as Tạ Quang Bửu and Ngụy Như Kon Tum in physics and Hoàng Xuân Hãn in mathematics.

The French war of reconquest broke out and the Vietnamese nationwide resistance began on the night of December 19, 1946. Citizens evacuated Hà Nội and went to the countryside. Professor Nguyễn Thúc Hào, together with his wife and three small children, walked dozens of kilometers to arrive in Ninh Bình where they boarded a train heading for Hào's native village in Nghệ An, 300 kilometers from Hà Nội. In early 1947, the Ministry of Education authorized him in opening a general mathematics class in his home village. At the same time, Professor Nguyễn Xiển began a similar class in the Northern Midlands (Vĩnh Yên Province) with a dozen students. However, during the French attack on Việt Bắc in October 1947, Nguyễn Xiển's students took refuge in caves and then returned to their home. They did not go to class until four years later when the resistance government decided to open the Teacher Training College and College of Fundamental Sciences.

Nguyễn Thúc Hào's "school" in Nam Đàn opened in the ancestor-worshipping house of one of his relatives, near Liễu Market. His first five students were teachers from Nguyễn Công Trứ Junior Secondary School in Vinh, who crossed the Lam River three times a week to attend the class. In the following years, the number of students increased to around twenty. Most of them had just completed senior secondary school. They lived at a house near the school, which was now located in another ancestor-worshipping house in Đan Nhiễm Village. In order to avoid detection and bombardment by the French, he used a coded address for correspondence: Mr. Nguyễn Tứ Đức, Nam Đàn Post Office.

"Why did I name myself Tứ Đức (Four Virtues)?" Nguyễn Thúc Hào later said, "Because my father was a successful Confucian scholar. He taught us that the four cardinal virtues for men are filial piety, brotherly affection, loyalty, and faith (*hiếu, đễ, trung, tín*). But they are also the four virtues advocated by Hồ Chí Minh: diligence, thrift, honesty, and righteousness (*cần, kiệm, liêm, chính*)."

In the midst of war, the final examinations of the College of Sciences were held in the free Interzone 4. The panel of judges included Nguyễn Thúc Hào, Đặng Phúc Thông, and Phạm Đình Ái, and was based at the temple dedicated to Mai Hắc Đế, a patriotic king, which was located at the foot of Đụn Mountain. Diplomas were granted only with the approval of the Ministry of Education, which was located in Việt Bắc in the North. In Interzone 5, a special examination was organized for a student, Hoàng Tụy, who had taken Nguyễn Thúc Hào's course through a correspondence. This student is now a world-renowned mathematician, the author of the Hoàng Tụy Theorem, Tụy's Cut, Tụy-Type Algorithms, Tụy's Inconsistency Condition, etc. Many other students of Professor Nguyễn Thúc Hào have also paid him homage with their remarkable accomplishments in a number of fields including science, education, and journalism.

During the last years of the resistance against French occupation, the Preparatory School for Higher Education and Teacher Training College were based in Interzone 4. Nguyễn Thúc Hào took part in managing these institutions together with renowned Vietnamese intellectuals at the time such as Trần Văn Giàu, Đặng Thai Mai, Đặng Xuân Thiều, and Nguyễn Mạnh Tường. The classroom was a communal house or a private house in the village. Classes were held at night-time. The students used the light of fireflies and mimeographed textbooks.

Peace was restored after the triumphal Điện Biên Phủ Operation in 1954. Professor Nguyễn Thúc Hào returned to Hà Nội to continue his work at the Science Teacher Training University where he became vice rector. The most difficult period in Nguyễn Thúc Hào's teaching career was the fifteen years when he served as rector of the newly-established Vinh Teacher Training University in Nghệ An. In the first five years, the university's structure was built. In the following seven years, it was constantly moved to and fro between Vinh and Thanh Hóa to avoid bombing. Then the university was reduced to rubble by American bombs and in the last three years before he retired, it had to be rebuilt.

Lê Văn Thiêm and Nguyễn Thúc Hào were among the first university mathematics professors accredited in the independent Việt Nam. They both were elected to be top leaders of the Việt Nam Mathematical Society. Professor Nguyễn Thúc Hào also took part in other social and political activities. He was a three-term member of the National Assembly, Vice-Chairman of the Fatherland Front of Nghệ An province, and a member of the executive committee of the Franco-Vietnamese Friendship Association.

Since his retirement in 1976, he has never stopped pursuing his studies. Apart from the two books already published, *Analytical Geometry* and *Differential Geometry*, he has completed three other books, namely *Mathematical Analysis*, *Vector Geometry* and *Linear Geometry*, which are waiting to go to press soon. Towards the end of his career, he translated fourteen mathematical works from Russian, English, and French into Vietnamese. At the age of over seventy, he went to Cambodia to teach at the Teacher Training College in Phnom Penh. He also lectured in the refresher courses run by the teacher training colleges in Huế and Vinh. He has always been a devoted scholar and teacher.

TRẦN ĐẠI NGHĨA – A PATRIOTIC ENGINEER

After the August 1945 Revolution the Vietnamese declared independence on September 2, 1945, French troops landed in Việt Nam in an attempt to reoccupy the country. After a preliminary agreement was signed on 6 March 1946, negotiations between the North Vietnamese government and the French were held at Fontainebleau Palace in France. On that occasion, President Hồ Chí Minh was invited by the French government to come to France as a distinguished guest.

Meeting Hồ Chí Minh

On June 26, 1946, a special French government airplane landed at Le Bourget Airport. President Hồ Chí Minh exited the aircraft. Minister of Overseas France Moutet, on behalf of French Prime Minister Bidault, solemnly received Hồ Chí Minh. He was also warmly welcomed by thousands of Vietnamese who were living in France, among them Professor Trần Đại Nghĩa. He was surprised by the President's bright eyes and simple dress, his modesty, pleasant manner, and kind face.

Several days later, through the Chairman of the Overseas Vietnamese Association in France, the President invited engineer Trần Đại Nghĩa to come and meet him in the Royal Monceau Hotel. He had the honor of companying the President to visit overseas Vietnamese organizations, the Palace of Versailles, the Normandy region (important World War II battlefields), and the Memorial to the Unknown Soldier.

"I liked Uncle Hồ's extraordinary simplicity, his scientific method of work, his humanity, tolerance, and presence of mind. He was able to cope with any situation," he recalled. "He was warmly welcomed everywhere by French workers who deeply respected him. He understood the psychology of all groups of people, including intellectuals. Once when Uncle Hồ and I were alone, he asked me about what I had been doing in France. I told him about my public activities (which everyone in the Vietnamese community knew) and also my secret activities (which only the Chairman of the Overseas Vietnamese Association knew). On September 8, 1946, he told me about the failure of the Fontainebleau Conference, saying "I am going home. Prepare to come with me. We will go in a few more days!" This was easy for me because I had prepared myself for this opportunity long ago. I had already given my notice, although at the time I was a chief engineer of an aircraft manufacturing company, having a good salary of 5,500 francs per month (equivalent to 22 taels of gold)."

On September 16, the French Government arranged a special trip by train from Paris to Toulon, which was reserved for President Hồ Chí Minh. Toulon is a port city on the Mediterranean Sea. Engineer Trần Đại Nghĩa accompanied the President and a group of other Vietnamese intellectuals such as Dr. Trần Hữu Tước and engineer Võ Quý Huân. Next morning, at 7 a.m., the train arrived in Lyon where it stopped for a long time. President Hồ Chí Minh and Trần Đại Nghĩa stayed on the train while others got off for a walk.

"At home, things are very difficult now. Can you endure the hardship?" Uncle Hồ asked.

"Yes, of course, Uncle," said he.

"We have no engineers or weapon builders and few machines as well. Can you work under those conditions?"

"I have been preparing for eleven years. I believe I can do it."

Years of Preparation

If it was not for his eleven years of preparation (from 1935 to 1946), engineer Nghĩa could not have contributed to the Anti-French Resistance War in Việt Nam in the way in which he did.

Trần Đại Nghĩa, whose real name was Phạm Quang Lễ, was born on September 13, 1913 in Xuân Hiệp Commune, Tam Bình District, Vĩnh Long Province (southern Việt Nam). His father, a schoolmaster who knew French and could read Chinese, often helped him with mathematics and taught him to remain faithful to the rites (meaning "lễ" in Vietnamese). His mother, who often chanted the name of Buddha, taught him to be gracious.

In the summer of 1933, in Sài Gòn, Trần Đại Nghĩa left the Lycée de Chasseloup Laubat with both French and Vietnamese diplomas, placing first in both of the examinations. However, he was unable to pursue his studies at the university in Hà Nội. His father's death left the family without resources. Two years later, in 1935, he was able to go to France for study thanks to a grant from a generous benefactor. While pursuing his studies in civil engineering, he managed to learn the technique of manufacturing weapons so that he could return to Việt Nam and contribute to the resistance against foreign invaders.

"This was not an easy task," he recounted later. "No country in the world published technical documents on their military. The French Empire, for good reason, did not admit Vietnamese students - even those who had been granted French citizenship - to schools that taught weaponry, military research institutes, or arms factories. That is why, during the eleven years, I had to study on my own in a silent and secret manner."

Among the thousands of Vietnamese students who came to France, only he had such an "odd" determination to study weapons. At that time, he could not imagine what Việt Nam's liberation war would be like. However, he believed that the Vietnamese people would rise against the foreign invaders, and then they would badly need weapons. He was confirmed in his determination by the historical example of Cao Thắng (1864-1893) in the Hương Khê Uprising led by patriotic scholar Phan Đình Phùng. Cao Thắng had instructed hundreds of blacksmiths to craft efficient guns copied from the model of 1874 muskets captured from the French. Trần Đại Nghĩa wanted to follow his example. He was very lonely in this circumstance. No one shared what seemed to be a strange extravagance. Only one of his friends agreed to help him in his search of military records which he read carefully.

In order to obtain a general view of the subject from top to bottom, he tried to understand the philosophy of war, strategy, tactics, and techniques of each military branch. He studied most types of weapons from Navy, Ground Forces, and Air Forces. He also read books on explosives, intelligence, and code. Even what seemed to be obsolete in Europe could be of use in Việt Nam.

First, he went to search through public libraries. Everyday, after his civil engineering lectures ended, he went to the Paris Public Library. On average, out of 20,000 titles, he found one work relating to armaments. Out of more than a million books in the library, only fifty were relevant.

Sometimes, he experienced difficult times. Homesickness and the memory of his dead father and his sister who died in a recent shipwreck would bring tears to his eyes. Several years had passed since he last received a letter from his mother and he did not know if she stayed in Vĩnh Long or left for Sài Gòn.

If one wants to have a thorough knowledge of military science and techniques of manufacturing weapons, one needs to grasp fundamental sciences – mathematics, physics, chemistry, and mechanics – as well as civil engineering such as mechanical engineering, electronics, and aeronautics. Therefore, he earned a bachelor's degree in science from the Sorbonne University, a diploma in civil engineering from the National College of Bridges and Highways, and an aeronautical diploma from the Institute of Aeronautics. He also received certificates from the Polytechnic University and the College of Mines.

In France, there are also the libraries reserved for professors. He tried to obtain admission to these secret libraries and acquire snippets of knowledge about armaments from his teachers. Through patience and skill, he succeeded in collecting over 30,000 pages of documents, most of which were secret. How did he thread his way through the vast and inextricable field of knowledge? Without teachers' instructions, how could he learn fundamental concepts? He had no opportunity to go to military research institutes or arms factories for practice. So, in his mind, he formulated hypotheses about the circumstances in Việt Nam and tried to find concrete solutions.

In September 1946, he returned home with President Hồ Chí Minh aboard the Aviso Dumont d'Urville, bringing along one ton of books. Dr. Trần Hữu Tước and some overseas Vietnamese accompanied them.

Return

His forty days aboard the ship with Uncle Hồ left indelible memories in his mind.

Trần Đại Nghĩa recalled the simple but emotional words by Uncle Hồ, "Our country is still poor as the result of eighty years under French exploitation and occupation. In the south of our country, blood is being shed. We have to fight for a long time until our happiness comes. You are now coming back to serve your country and your compatriots."

Hồ Chí Minh stood up to the enemy's provocations without flinching. On the open sea, a naval maneuver was launched under the orders of the French captain. Cannons, anti-aircraft artillery, and machine guns made thunderous noises. While his attendants felt anxious, Hồ Chí Minh quietly smoked a cigarette and smiled. "They are testing your courage!" he said.

Three times a day, Trần Đại Nghĩa and the other returnees gathered around Uncle Hồ to discuss internal affairs and international events. Uncle Hồ also asked them to try and win the sympathy of the sailors. Naturally Trần Đại Nghĩa was confused since he had never done any mass organizing.

On October 20, 1946, the Aviso Dumont d'Urville landed at Hải Phòng. The next morning, Trần Đại Nghĩa met Võ Nguyên Giáp in Hà Nội, at that time in his early thirties. Võ Nguyên Giáp warmly shook his hand and said, "We have been waiting for you. We were very glad to learn that you were coming back with Uncle Hồ."

He then went to Thái Nguyên to meet Tạ Quang Bửu and examine American bazooka shells to find out how to manufacture them. Việt Nam did not have the special explosives used in these shells. He proposed the use black powder explosives and the preparation of a suitable cartridge. During the first test, the shell exploded violently, but the projectile was unable to pierce through the steel. It turned out that the workers who made the cartridge had given it a thickness of 4 millimeters instead of 1-1.5 millimeters as planned.

On December 5, 1946, he went from Thái Nguyên to Hà Nội and was invited to meet Uncle Hồ who said, "The work assigned to you is for the sake of the people, so I give you the alias 'Trần Đại Nghĩa' (Great Cause for Justice). The alias will also help you protect yourself, your family, and your relatives who are in the South. 'Trần' is borrowed from the thirteenth-century general Trần Hưng Đạo; 'Đại Nghĩa' appears in *Bình Ngô đại cáo* (Great Proclamation upon the Pacification of the Wu) by the fifteenth-century famous poet Nguyễn Trãi. Do you like your alias?"

Trần Đại Nghĩa was deeply moved by the fact that although Uncle Hồ was very busy with state affairs, he spent time considering a name for him. On that day, Trần Đại Nghĩa was also appointed to the post of Head of the Department of Weapons.

Half a month later, the national resistance broke out on the night of December 19, 1946.

Participating in the War

In a hurry to leave Hà Nội, Trần Đại Nghĩa was able to carry with him only a few dozen kilograms of the ton of books and documents on armaments that he had brought back from Paris.

Fighting raged in the streets of Hà Nội. He and his staff had to urgently produce Molotov cocktails and Japanese-style anti-tank mines so that the suicide troops could use them to destroy the French tanks. He designed new models of grenade launchers and 50.8-mm mortars. Copper, rather than cast iron, was required to make mortar shells, but no copper could be purchased in the mountainous regions. Therefore, military workers asked for help from local residents who responded enthusiastically to the armament effort. They brought all their copper objects to the military forge: pots, trays, incense-burners from family altars, and temple bells. He was stunned to see these piles of gifts. He refused to accept worship objects such as incense-burners and bells and only kept other utensils although he knew that these were highly valued by the peasants.

Early in the resistance war, the Việt Minh captured an enemy munitions depot in Vĩnh Yên containing five hundred

300-mm naval cannon shells. Soldiers immediately moved the shells to Tuyên Quang where they were unscrewed and emptied of explosives – a very dangerous task. One man was killed instantly, pieces of his body was scattered all over a sweet-potato field. However, the workers did not give up and finally obtained sixty tons of explosives to make bazooka shells. Theoretically, the bazooka is an anti-tank weapon but it was used by Việt Minh soldiers in all sorts of operations: destroying a nest of machine guns at close range, blowing up enemy troops camped inside a building, or shooting at a riverboat from a hiding position in reeds along the bank.

The arms workshops ran day and night. With steel taken from railroads, the workers succeeded in making all the pieces of bazookas and even the 50.8-mm mortar. In a workshop in northern Interzone 3, which had no machines but only long anvils, everything was made by hand. Yet the arms produced there lacked nothing in precision.

He worked together with soldiers in the arms workshops located in the forest to contribute to the resistance war.

Creation

He creatively applied his experience to the circumstances of the resistance war. From the beginning, he rejected the idea of making high-quality explosives like trinitrotoluene (TNT) because Việt Nam lacked materials and equipment. The small quantity of TNT taken from the French was reserved for making bazooka rockets, long-range mortar shells, and top-grade mines. For other weapons, only black powder-clorat – a mixture of saltpeter, sulfur, charcoal and kali clorat – was used. The method of producing saltpeter used by foreigners proved too complicated for Việt Nam's conditions. The

Việt Minh resorted to the traditional method of mountain dwellers: climbing to the roof of caves with rope ladders to scrape off bat excrement which was weathered to become saltpeter. Amidst the forests, the Việt Minh had learned to produce sulfuric acid through traditional methods.

In general, in the world, there are two types of grenades: offensive grenades with sheet-iron casing, efficient at a range of fifteen meters, and defensive grenades with cast-iron casing and a range of action of fifty meters. However, due to the lack of good explosives and the imprecise specification of materials, the Việt Minh had to make 50-60 different types of grenades.

When the enemy built fortified battalion positions, the 81-mm mortars of the Việt Minh were too small to threaten them. He discussed the problem with his colleagues, pointing out the need for larger-gauge steel tubes. One officer solved the problem by discovering 205-mm steel tubes. One worker told them that there were oxygen containers in the area. He calculated that these would be sufficient and went to work. After several months' time, the troops had 205-mm mortars at their disposal.

In manufacturing weapons, one small careless mistake made by an engineer could lead to the user's death. Therefore, specifications of materials used for making weapons had to strictly comply with the requirements. However, in the resistance war, there were not materials with precise specifications. It was a major difficulty for him, and he could only overcome it through creativity.

After the 1947 Winter-Spring Campaign, Trần Đại Nghĩa, who was then thirty-five, was promoted to Major

General. He dreamed of making a light, portable weapon but with the destructive force equal to a cannon made from six tons of steel. He thought of the recoilless gun (SKZ) used for the first time by the Americans in the battle of Okinawa (Japan), which he heard of when he was in Paris. Since that time he had tried to understand the principle of this modern weapon. Now, amidst the Việt Bắc forest, with no technical reference documents, he decided to try and build a recoilless cannon or ĐKZ. He was supported by Minister of Defense General Võ Nguyên Giáp and Vice Minister of Defense Tạ Quang Bửu.

"My collaborators and I had to work for fourteen hours a day on the new weapon," he later recalled. "We needed total concentration in order to solve difficult problems. A scientific researcher is not unlike a blacksmith. When the iron is red-hot and malleable, you must hammer it quickly without giving it time to cool down. My mind was very tense. Designing the ĐKZ obsessed me all the time, not only in my work hours, but also when I went out for a walk, when I took a bath in a mountain stream, or when I ate or slept. Sometimes, no ideas came to my mind in the day but they sometimes came when I was in bed. I always had a piece of paper and a pencil near my bed within reach and I noted down any thought that came at any moment, even at midnight, before it vanished."

He was surrounded by a team of enthusiastic young scientists such as Hoàng Đình Phu, Nguyễn Trinh Tiếp, Bùi Minh Tiêu, Phạm Đồng Điện, Nguyễn Văn Hường, and others.

The ĐKZ first appeared in the battle of Phố Lu, destroying the enemy's blockhouse. With a barrel fifty millimeters in diameter alongside a cannonball with a diameter of 160 millimeters (three times more than that of barrel), the weapon looked quite strange. The handle of cannonball was made of wood – a Vietnamese innovation. After sustaining direct fire from a distance of 100 meters, the assault took place immediately. The Command of the French Indochina Army was appalled by the ĐKZ and gave a secret order to assassinate Trần Đại Nghĩa.

In 1950, troops on the south-central battlefields received 10 ĐKZ and 150 cannonballs which were able to penetrate concrete walls to a thickness of several meters. One night, the Việt Minh destroyed five enemy posts, forcing the disorderly retreat of enemy troops from hundreds of posts. Soon afterwards, the ĐKZ was sent to Trần Đại Nghĩa's home region in the Mekong River Delta.

After the ĐKZ, he next thought of manufacturing rockets. Two difficulties appeared at once: the explosive force of the black powder used by the Việt Minh was too weak, and lacking steel, the copper tubes they produced exploded too easily under pressure. To make matters worse, he had lost all his documents in 1947 during a French paratroopers' attack on Bắc Cạn. Relying entirely on his memory, he was able to design an effective small rocket weighing thirty kilograms and capable of reaching a distance of four kilometers. He was the first intellectual who was given the title "Hero of Labor" at the National Congress of Heroes and Model Workers (held in Việt Bắc in 1952) because of his achievements in manufacturing weapons.

Professor Trần Đại Nghĩa recalled, "In the first years of the anti-French war, Uncle Hồ sent me quarterly letters, encouraging and reminding me. I remember a piece of his advice, 'We must defeat the enemy while limiting our casualties as much as possible.' When I lived in Paris, I was addicted to smoking since I worked an average of fourteen hours often alone in a single room. Uncle Hồ knew about this bad habit of mine and ordered the General Department of Logistics to get me a pack of Cotab cigarettes every day. These good-quality cigarettes were only available on the black market in the occupied zone. Uncle Hồ said, 'Without cigarettes, our poor Nghĩa's mind goes numb.' In Việt Bắc, he always reminded me, 'Anytime you have a problem, don't hesitate to come and see me!' Each time I asked him for an appointment, I got a horse ready for the journey because I knew I could go immediately."

In 1954, the glorious victory of Điện Biên Phủ put an end to the resistance against the French.

Nine years later, in February 1965, American warships and planes began to bombard North Việt Nam. Trần Đại Nghĩa was summoned to the Presidential Palace.

"According to you, what is the best way to limit the victims to a minimum?" Hồ Chí Minh asked him.

"In World War II, the Soviet Union lost twenty-two million people, most of them civilians, because civil defense was not carried out thoroughly, meaning the populace and the civil service were not evacuated from the cities. In Germany, the American bombing of Hamburg killed hundreds of thousands of people, mostly civilians. European city-dwellers, accustomed to living with the comfort of water and electricity, were afraid of leaving town," he answered.

"We will carry out a systematic evacuation. But we must also have people to hold the cities."

"Of course, Uncle! But we must build underground shelters for these people, both collective and individual. Even cyclists must be able to jump down in an individual bunker as soon as the alarm siren is heard."

"That's a good idea. I'll have it carried out at once."

On the instruction of Uncle Hồ, several days later Generals Võ Nguyên Giáp and Nguyễn Chí Thanh proposed Trần Đại Nghĩa as Vice Head of the General Department of Logistics, responsible for military technique. At that time, he was already serving as Chairman of the State Committee of Sciences and Technology and had been elected a member of the Mechanics Committee of the Academy of Sciences of the Soviet Union.

The American government mobilized as many as 5,000 doctoral scholars to develop weapons for use in Việt Nam. On the Vietnamese side, under the direction of Professor Trần Đại Nghĩa, there were only a few small groups of scientists and technicians, few of them having a doctorate degree. Yet they succeeded in improving B-52 equipment to interfere with radar waves and increase the range of SAM-2 rockets to reach the height of a B-52 plane. Therefore, the Vietnamese army was able to cope with American B-52 strategic bombers and MK-52 torpedoes. This occurred due to the close cooperation between the High Command of Air Defense, civil engineers, naval forces, the Institute of Military Technology, and researchers of universities in Hà Nội.

Trần Đại Nghĩa received the Hồ Chí Minh Award for his research on and the manufacture of rockets, recoilless guns, and missiles during the anti-French resistance (1945-1954).

"For those of our generation, the life and work of academician Trần Đại Nghĩa are legendary," said Nguyễn

Văn Hiệu. The former rector of Vietnam National University (Hanoi), Professor Nguyễn Văn Đạo, a famous mechanician, said that his generation "regarded Trần Đại Nghĩa as a beloved father and brother."

Trần Đại Nghĩa was the first rector of Hà Nội Polytechnic University.

On his birthday, with tears in his eyes, Professor Trần Đại Nghĩa spoke to his guests, "Uncle Hồ gave me my determination. Thanks to his teaching, I could accomplish my task. His image will forever be in my heart. I owe him deep gratitude. He helped me overcome difficulties and trials. It allowed me to bypass the misunderstandings which sometimes my superiors and my colleagues showed me."

LÊ VĂN THIÊM – A PATRIOTIC MATHEMATICIAN

Until French colonization at the end of the nineteenth century, Vietnamese traditional education was based on humanities. It did not teach the hard sciences, particularly mathematics. *Đại thành toán pháp* (Synthesis Knowledge about Mathematics) written by Lương Thế Vinh in the fifteenth century was the only important work in the field.

Under the French colonial regime (1884-1945), a handful of Vietnamese intellectuals studied mathematics in France. These included Nguyễn Xiển, Hoàng Xuân Hãn (the first "agrégé" - equivalent to the Master degree - of mathematics), Tạ Quang Bửu, and Nguyễn Thúc Hào. When they returned to Việt Nam, they taught at secondary schools and colleges.

Lê Văn Thiêm was a truly gifted mathematician. He was the first Vietnamese to successfully defend a doctoral thesis in Mathematics in Germany (1944) and a State Doctorate (doctorat d'Etat) thesis in mathematics in France (1948). He taught mathematics and mechanics at Zurich Polytechnic University (1949), where Albert Einstein had pursued his studies.

Lê Văn Thiêm founded the Institute of Mathematics in Việt Nam in 1959. The staff was initially consisted of one professor and three mathematicians with a doctoral degree, and twelve teachers with a bachelor's degree. In 2001, it consisted of fifteen professors, thirty PhDs, and forty-two doctors of science. The institute has gained a nationwide and worldwide reputation thanks to its influential research works.

Lê Văn Thiêm was born in 1918 in Trung Lễ Village, Đức Thọ District (Hà Tĩnh Province), a poor village famous for its tradition in the struggle against French colonialism. His family lineage, the Lê, is celebrated for a great number of laureates of the triennial royal exams and for the participation in the movement Cần Vương (Serve the King), fighting against French domination. In particular, Lê Thước, his brother, was an erudite professor who had thorough knowledge about Sino-Vietnamese and French cultures. When the August 1945 Revolution broke out putting an end to the 80-year French domination, Lê Thước and his son, Lê Thiệu Huy, immediately joined Hồ Chí Minh's cause. Lê Thiệu Huy, a young talented scientist, became chief-of-staff of the Lao-Vietnamese joint troops.

Pursued by the enemy, he crossed the Mekong River in a boat together with Prince Souphanouvong, leader of the Lao Resistance Movement. He covered the prince with his body against the enemy's fire and was killed. He was twenty-six years old at the time. The heroic sacrifice of his nephew, who was two years younger, enormously impressed Lê Văn Thiêm. Many years later, he dedicated his doctoral thesis on mathematics to Huy, "To my beloved friend Lê Thiệu Huy who sacrificed his life to the Vietnamese motherland on the Mekong River in spring 1946."

When Lê Văn Thiêm was born, he was given the popular name of Thêm (which means "additional" or "extra") because his parents had not expected such a birth, thinking that his elder brother would be their last child. In Sino-Vietnamese translation, Thêm is pronounced as Thiêm. His family was very poor. When he was six or seven years old, he had to tend his family's buffalo. He learned the Romanized Vietnamese script at home with his ninth brother. After his mother died when he was thirteen years old, he lived in Quy Nhơn with his eldest brother, Lê Văn Kỷ, who was then an "Indochinese physician"[1] in this town.

Lê Văn Thiêm's mathematical ability surprised Michel Casimir, BSc of Paris and Rector of Quy Nhơn National School. The latter said, "He will go further than me."

This prediction became reality.

After he finished the first part of his education (equivalent to the eleventh grade of today's senior secondary school), he easily attained the first level of Baccalaureate (GCE). The next year, he passed the GCE exam. At the time, the Indochinese University had only two faculties: law (for training government officials) and medicine. However, he was not interested in either of these fields. Therefore, he enrolled in the Preparatory School for PCB (Physics-Chemistry-Biology). In 1939, he was rated second in the PBC final exam and obtained a scholarship to go France for further study. It was the last chance for Indochinese students to go abroad because in September 1939 World War II broke out.

1. A graduate of the Hà Nội Faculty of Medicine, Lê Văn Kỷ took part in the last royal civil service exam in Huế and did well. However, he did not want to become a mandarin of the puppet court in the service of the French colonialists, so he devoted himself to medicine.

In Paris, he passed the entrance exam into École Normale Suprérieure of Ulm Street. His schoolmates during the 1941-1942 school year were well-known mathematicians such as Max Fouvieille, Robert Carol, General Debreu, and Paul Roussel. According to his friends, he was sweet and modest. They were very surprised that in his country, local intellectuals could only occupy subordinate positions in the colonial machine. After obtaining a BSc degree, he worked in Switzerland with Nevanlinna on meromorphic functions, which gave him some ideas on his future thesis.

At the end of 1942, when he came back to Paris, the situation worsened with the Nazi occupation. Together with some young Vietnamese, he asked for permission to go and study civil engineering in Germany so as to serve Việt Nam in the future. Their request was rejected because Berlin found that these disciplines were too close to the military field. At last, he found a German professor who undertook to supervise his Doctorate B thesis (i.e. Doctorate Habil) due to his supervisor's health. In Berlin, he had opportunity to witness how the Nazi regime was overthrown.

Back in France in 1948, he successfully defended his State Doctorate thesis under the supervision of Georges Valirơn, a great specialist in analytical function of a complex variable. Lê Văn Thiêm has been the first mathematician to solve the problem posed by Nevanlinna: an inverse problem in the meromorphic function theory. He had suggested a new orientation for promoting Nevanlinna's theory.

In 1981, Lê Văn Thiêm and two other Vietnamese mathematicians, Hoàng Tụy and Nguyễn Văn Đạo, were invited to spend a month studying in the United States. A young American mathematician thanked Lê Văn Thiêm for his doctoral thesis published in 1944 in Germany, which had given the former his topic for his doctoral thesis.

In 1949, Lê Văn Thiêm taught at the Zurich Polytechnic University. However, he did not feel at peace in the beautiful city of Switzerland. The death of his nephew, Lê Thiệu Huy, on the Mekong River reminded him of his debt to his nation.

In 1949, refusing a lucrative position in the United States, he bought an air ticket to Bangkok with his savings. He crossed the Cambodian jungle, which was then controlled by French troops, to join the South Vietnamese resistance base.

In May 1950, he came to the northern highlands where the resistance government was based. Walking over mountains and through forests was a real expedition. It took him nearly a year to cover the 2,500-km trail which would be later called Hồ Chí Minh Trail. The headquarters of the Medical University and the nascent colleges were located in Chiêm Hóa, Tuyên Quang Province. He was very happy to meet Dr. Hồ Đắc Di, General Director of the Higher Education Commission and Rector of the Medical University as well as other distinguished intellectuals, such as surgeon Tôn Thất Tùng, specialist in malaria Đặng Văn Ngữ who had just returned from Japan, and Minister of Education Nguyễn Văn Huyên. Lê Văn Thiêm was nominated as Director of the College of Fundamental Sciences and Central Teacher Training College.

He met Deputy Minister of Defense Tạ Quang Bửu in a jungle hut. He was very pleased to find out that his senior by eight years, though doing multiple tasks, did not give up his love for science, as shown by heaps of French, English, and German books and journals which littered his office.

In 1939, when Lê Văn Thiêm arrived in Paris, Tạ Quang Bửu had gone back to Huế, where he would teach at a private secondary school. In 1946, Tạ Quang Bửu came back to France as a member of the Vietnamese delegation to the Fontainebleau Conference. He did not forget to pay a visit to his compatriot from Nghệ Tĩnh.

At the National Conference on Culture held in Việt Bắc in 1948, while Lê Văn Thiêm was not yet back in Việt Nam, Tạ Quang Bửu informed the audience of Lê Văn Thiêm's State Doctorate thesis, saying, "The author of this thesis, a young man of thirty, ranks among our masters."

The close cooperation between the two eminent mathematicians constituted an important factor and a guarantee for the building of mathematical sciences in Việt Nam.

At the end of 1951, with the permission of the Chinese government, the College of Fundamental Sciences and the Central Teaching Training College were moved to Chinese territories to avoid French bombings and land attacks. They were based in the Educational Quarter located in the suburb of Nanning, Guangxi Province. The teaching staff of the two schools initially consisted of nine persons, three of whom taught mathematics. They included Lê Văn Thiêm, Nguyễn Xiển, and Nguyễn Cảnh Toàn (assistant). The textbooks were the *Mathematical Physics Method* by Courant Hilbert and some other publications brought by Lê Văn Thiêm from Switzerland. In Nanning, one could buy Russian books on mathematics, for instance, *Geometry* by Lobachevski. To read it, Nguyễn Cảnh Toàn had to consult a dictionary very often. Meanwhile, in Thanh Hóa province, the Ministry of Education opened a Preparatory

School for Higher Education and a Teacher Training College for Interzone 4.

In May 1945, the Điện Biên Phủ victory put an end to the war. The resistance government established a commission in charge of taking over the higher education institutions in Hà Nội occupied by the French. Lê Văn Thiêm took part in this commission.

Hà Nội was liberated on October 10, 1954. Lê Văn Thiêm was nominated Director of Hà Nội Teacher Training College of Sciences which would train distinguished scientists such as Nguyễn Văn Hiệu, Nguyễn Văn Đạo, Vũ Đình Cự, and Phan Đình Diệu.

In 1956, Hà Nội University was founded by Lê Văn Thiêm as Vice-Rector and Dean of the Mathematics Department. Being the founder of the Việt Nam Mathematical Society, he was also its first chairman. He assumed other important responsibilities: he was the deputy chairman of the National Assembly, the plenipotentiary delegate of Việt Nam to the Dubna Joint Institute for Nuclear Research, and a member of the Việt Nam State Committee on Science and Technique.

Together with his students, Lê Văn Thiêm often tried to apply mathematics to concrete problems of the country such as agricultural irrigation, prevention of floods, and other natural calamities.

In 1963, he invited Soviet academician M. A. Lavrentiev to make a report in Việt Nam on the application of the function of a complex variable to the theory of explosions.

Then, in collaboration with his students, he applied the Lavrentiev method to some concrete problems in Việt

Nam. These included calculations to determine the way to use the Núi Voi stone quarry for the building of the Thái Nguyên Metallurgical Plant, cooperation with the Technical Department of the Defense Ministry in the mining for road construction (1966), and cooperation with the Ministry of Transport and Communications in the mining for dredging the section of the Nhà Lê canal stretching from Thanh Hóa to Hà Tĩnh (1966-1967).

After the Institute of Mathematics was founded, he realized that the functions of a complex variable should be applied to other fields, such as the elasticity theory and the movement of viscous liquid. He and his collaborators, including Ngô Văn Lược, Hoàng Đình Dung, and Lê Văn Thành, also resolved other problems. He proposed an original solution to the problem of infiltration through two-stratum oil, using the principle of axial symmetrical extension of analytical function. This application is highly valued and quoted in the book *The Theory of Groundwater Movement* written by Russian academician P. Ya. Polubarinova Kochina (published in Moscow in 1977).

He also applied the analytical function theory to solving the problem of movement of vicious liquids. The result of his research was presented in a report made by Lê Văn Thiêm himself to the International Congress of Mathematicians held in Vancouver (Canada) in 1974.

Having suffered diabetes and high blood pressure for a long time, Lê Văn Thiêm died on July 3, 1991 in Hồ Chí Minh City where he had lived and worked in the last years of his life.

At the time of his death, late Prime Minister Premier Phạm Văn Đồng wrote to Lê Văn Thiêm's wife, "His death has for all the better brought out the stature of a mathematician and a dutiful communist and contributions made by him. I say this to you from the bottom of my heart, and to his soul and to the whole world".

In Paris, academician Laurent Schwartz, winner of the Fields Medal (equivalent to the Nobel Prize), wrote a long article for the Bulletin of the Alumni of École Normale Supérieure in Paris. In the article, he mentions Lê Văn Thiêm's professional activities, laying stress on his efforts to train a pool of Vietnamese mathematicians. He emphasizes that, thanks to Lê Văn Thiêm's contributions, Việt Nam has become a country achieving a high level of research in mathematics in the Far East. He says that Tạ Quang Bửu and Lê Văn Thiêm waged a difficult struggle for scientific impartiality. Lê Văn Thiêm is regarded by his Vietnamese colleagues and his old friends of École Normale Supérieure in Paris as a kind-hearted and modest man.

Professor Nguyễn Văn Đạo also dedicated an article to Lê Văn Thiêm, which is published in a special issue of the *Journal of Mathematics* (published by the Vietnamese Academy of Science and Technology). Following is a short quotation, "What is precious about Lê Văn Thiêm is that his life was governed by the truth, which was applied to all his thoughts and speeches. He judges his students and colleagues impartially and openly. His views on humans and life are clear and open. In his personality is altruism, a trust in man's goodness, a hope for a better life for everyone, and a belief in the progress of our science. In the universe there exist dead stars that go on shining. Lê Văn Thiêm is one of these stars in the scientific firmament of Việt Nam."

ĐỖ TẤT LỢI – A SCIENTIST DEVOTED TO TRADITIONAL MEDICINE

In 1945, a month before the August Revolution which would put an end to eighty years of French colonization, Hồ Chí Minh was severely ill. However, he was saved from death thanks to a recipe of natural medicines. He had been living since June in Tân Trào Village, 150 km north of Hà Nội. In this northern mountain zone liberated by the Việt Minh, he was preparing the final stage of the liberation movement.

Uncle Hồ fell ill just at the moment when the situation was swiftly changing. The Central Committee and the National Assembly were going to convene at Tân Trào. He had a high temperature and was emaciated and weak. He was delirious when General Võ Nguyên Giáp came to see him. The General made him take aspirin and quinine, but they had no effects. During the night, in a hut at Nà Lừa, Uncle Hồ talked about taking advantage of the current favorable circumstances in order to seize power. Thinking he was at death's door, he wanted to write his will. The next morning the General asked the ethnic Tày villagers whether there was a good medicine man in the region. As soon as one was found, he was invited to come at once. The old man felt Uncle Hồ's pulse and burned wild roots which were then mixed into a soup and administered. By midnight, Hồ Chí Minh had begun feeling better and was restored to health in a few days' time.

Ten years later, in 1955, President Hồ Chí Minh wrote to the Congress of Public Health, "Our ancestors made a great number of experiments on the treatment of diseases with Vietnamese and Chinese remedies. Dear nephews and nieces, in order to enlarge the field of medical knowledge, you must investigate Eastern medicine and combine it with Western medicine".

This point of view was also adopted by Prime Minister Phạm Văn Đồng, "Most of our health cadres, trained by Western universities, have not realized that for millennia, our people, without the help of Western medicine, had excellent recipes, efficient medicines, and medical and pharmaceutical traditions based on experience of long standing.... If we study traditional medicine and study its legacy, we are sure to be able to quickly solve medicinal and pharmaceutical problems in a simple, cheap and original way." ("Address to the Emulation Congress of Public Health," 1965-1966)

A Short History of Traditional Medicine

Far from being a mere copy of Chinese medicine, Vietnamese traditional medicine has two components: southern medicine, which is entirely Vietnamese, and northern medicine, which is Sino-Vietnamese.

The former component comes from communities of fifty-four ethnicities living in the southern area. Primitive recipes made use of tropical fauna and flora in their raw state. The two fathers of traditional medicine, Tuệ Tĩnh (seventeenth century) and Lãn Ông (eighteenth century), although trained at Chinese schools, saw southern medicine as a valuable heritage to be combined with the traditional medicine in the North.

Northern medicine dates back to 1800 B.C. More theoretical, it forms a well-articulated system and relies on a rich medical literature.

Vietnamese traditional medicine has ensured the provision of healthcare for the majority of the people for thousands of years, serving as a general set of conventional concepts for an empirical yet efficient treatment of diseases.

Under French domination, the traditional medical system was despised. The Faculty of Medicine and Pharmacy of French Indochina only trained a small number of students. Therefore, modern medicine was in the service of the most favored classes of city-dwellers. Students learned the names of medicinal plants and *materia medica* in Latin and French. A large number of medicinal substances (such as maize silk, pomegranate skin, tea, and strychnine) were exported to France and then brought back to Việt Nam in the form of patented medicines.

On the other hand, according to Western doctors, anarchy reigned in traditional medicine. There was no specialized school: Confucian scholars studied by reading books and each physician had two or three disciples. Medicinal substances were imported indiscriminately from China, some of them existing already in Việt Nam.

In July 1943, the Governor-General of French Indochina issued a decree forbidding the use of certain traditional medical substances considered to be poisonous. This was a serious blow to traditional medicine. However, the 1945 Revolution brought a renewal of interest and hope.

During the anti-French resistance (1946-1954)

On October 31, 1946, an article appeared in a Hà Nội daily newspaper defending traditional medicine and pharmacies. The essay was written by a 27-year-old pharmacist just out of the Hà Nội Faculty of Medicine and Pharmacy, Đỗ Tất Lợi. He wrote, "Our northern medicine and southern medicine, accumulating thousands of years of experience, are on the decline. To save them - to save the precious heritage of our ancestors - we think they must be renovated. When we have pharmaceutical chemists, masters of the Western scientific method and also of traditional medicine, traditional medicines will have good chances to develop.

If we apply science to the preparation of medicine and the infusion of herbs, we can create solid assets and planted solid landmarks to define the field of research in our traditional medicine. Then traditional medicine will once again retain its former importance. It then can be combined with Western medicine because we will have found the constituent principles of the medicinal substances of our traditional medicine."

This profession of faith served Đỗ Tất Lợi for his entire life. During his years of study at the Faculty of Medicine and Pharmacy (1939-1944), he had wanted to discover the secrets of traditional medicine and pharmacy.

Could this passionate interest have come from his childhood? He was born in 1919 in Phù Xá Village (Vĩnh Phúc Province). His father liked planting custard-apple plants and rearing deer, which no doubt inspired in him a love of nature. At the age of six, he suffered a bone fracture and was cured by a famous old healer named Sáp. Later, when he was a student, he did not rest during school holidays. He began learning the basics of traditional medicine and traditional medicine pharmacies with the healer. He continued expanding his knowledge in this field. After graduating from his university, he became acquainted with herb-sellers when he lived in Hữu Tiệp Village (Hà Nội). Through this he learned many popular recipes and curative properties of plants, some of which he could observe in gardens. He opened a pharmacy in Hà Nội and did not hesitate in preparing medicines with local plants. When the anti-French resistance war broke out, he joined the resistance at the beginning, riding out of Hà Nội on his bicycle with a few personal belongings. He was appointed head of the newly-created Institute of Research and Production of Medicines by the Military Department.

Fond of field research, he continually scoured the most remote corners of the mountains during the war to collect medical ingredients and plants. He devoted much of his time to talking with traditional healers in their stilt houses and herb-sellers in the market.

It is said that Hồ Chí Minh, in his stay in the jungle, advised his attendants to keep a provision of wild *mã lìn ón* – dried leaves and roots – in their traveling-bags as a remedies against high fever.

After investigations on this medicine commonly used by mountaineers, he finally identified this plant: it is the white *hà thủ ô*. This tropical liana, Streptocaulon juventas Merr., contains toxic alkaloids, grows in great number on the bare hills of the highland and midland regions of the North. At the same time, he also found *mã tiền* (Strychnos nux vomica Linn.), which was considered until then by French botanists as a plant able to grow only in the south of the country.

In 1948, Đỗ Tất Lợi and Nguyễn Văn Đàn published their work on chlorophyll, considering it a modern antibiotic. Later on, Đỗ Tất Lợi and his group were able to extract this substance from the leaves of bamboo and jujube trees to treat soldiers' burns and wounds.

Malaria was a plague that was decimating the soldiers stationed in forests. In order to fight the disease, he and his institute invented a new traditional medicine. To satisfy the needs of antiseptic washing and bandaging in military infirmaries, he investigated a long-standing traditional practice. With his colleagues, he extracted the eucalyptus oil, which can serve as an antiseptic in the treatment of burns and wounds.

Returning to Hà Nội

After the Điện Biên Phủ victory in 1954, Đỗ Tất Lợi came back to the liberated capital. Discharged from the army, he was appointed Director of the Department of Materia Medica and Botany of the Hà Nội University of Medicine and Pharmacy, and later Director of the Department of Materia Medica at the Hà Nội Pharmaceutical University.

At that time, he had been enjoying favorable circumstances in carrying out his work as a researcher and practitioner of traditional Vietnamese medicine. However, a colossal task was still waiting for him. In textbooks, medicines, chiefly imported from France, were registered under French and

Latin names. It was necessary to find an encyclopedia on Vietnamese pharmaceuticals. This would win the patients' confidence in Vietnamese medical practices.

In this work, everything was lacking. There were not enough competent cadres or financial means in the newly-liberated colonial country, devastated by nine years' resistance and threatened by American intervention. Some specialists even thought that the French authors had already discovered everything about botany, and there was nothing to do but read them. It is true that the French botanists had done good work, but Vietnamese traditional medicine remained an immense field with many possibilities for discovery.

Theoretical research and applications

Deprived of essential resources for research on traditional medicine, Đỗ Tất Lợi used any means that came to his land, solving problems in the best way possible.

He and his students interviewed herb-sellers, healers, and traditional practitioners, who often kept their recipes secret. To create a systematic and analytic inventory of medicinal plants of the country, they included every source of information.

A year after the liberation of Hà Nội, he requested the old traditional practitioners of Hà Nội to plant the medicinal plants that they considered the most precious in the uncultivated garden of the Temple of Medicine. The temple had become an appropriate place for professional exchanges in order to gain better knowledge of medicinal flora. After this fruitful experience, he proposed duplicating gardens of medicinal plants throughout the country.

In 1954, the Ministry of Domestic Trade found the General Company of Forest Products and Regional Specialties. It turned to his appraisal for the choice of materia medica, including medicinal plants, for new products proposed by different regions of the country.

He scoured the whole country in search of new medicinal plants and forgotten secret recipes. In the mountainous regions of the North, he found *kim anh* (Losa laevigata) in Lạng Sơn, *tục đoạn* (Dipsacus japonicas), and *đảng sâm* (Codonopsis sp.) in Lào Cai, *gấu tàu* (Aconitum fortune), and *hoàng liên* (Coptis) in Sa Pa. This materia medica had often been first exported to China and then imported back to Việt Nam in the form of patent medicines sold at exorbitant prices and labeled "Made in China."

It was by chance and also through his own patience that he discovered *cây ba gạc* (Rauwolfia) mentioned by H. Lecomte in his *General Botany of Indochina*. Five years after the liberation of Hà Nội, he succeeded in locating the anti-hypotension medicine. In Sa Pa, exhausted after long hours of mountain climbing, he met a young buffalo-tending boy who, after having listened to the detailed description of the plant, showed him a sample of it.

Thanks to two local schoolboys, he also found *hoàng tinh* (Polygonatum), used as a tonic, and other medicinal plants.

In his conversation with the common people he could not speak in a straightforward manner but had to resort to roundabout ways to reach his aim. For example, he would describe the symptoms of a disease before asking people what local plant was used to cure it.

Once, a foreign pharmacologist claimed, "The Vietnamese live on a heap of medical supplies, but still die of illness!" Through their work, he and his colleagues have tried to prove the second part of this statement as false.

Hundreds of tons of mint and menthol oil were imported every year for pharmaceutical and food industries even though mint grows naturally all over Việt Nam. He proposed mass cultivation of mint. The Ministry of Foreign Trade and the State Planning Committee worked with this process. He and his colleagues tried to acclimatized strains of mint from France, the USSR and China. Finally, the local species proved to be the best because it was already adapted to the country's conditions. Many cooperatives began to plant mint. Whereas in 1972 North Việt Nam produced only eight liters of mint oil, by 1976 the reunified Việt Nam produced tons.

He also contributed to the extraction of nerionline from the leaves of *trúc đào* (oleander). Noticing that most heart remedies were prepared from plants growing in foreign countries, he investigated the oleander growing on Vietnamese soil. After many experiments and clinical tests, he found that the Vietnamese oleander showed convincing therapeutic effects. He replaced the expensive and complicated Soviet method of preparation with a simpler and cheaper method.

Other achievements include the use of the flowers of *cây hòe* (Sophora) to prepare vitamin P, using *ngải* (Artemisia vulgaris Lin.) and *ích mẫu* (Leonurus heterophyllus Sweet) for gynecological purposes, the use of *ké đầu ngựa* (Xanthium strumarium Lin) for preparing a medicine against the goiter, and *nha đảm tử* (Brucea javancia) for fighting amoebic dysentery. He also scientifically studied the use of geckos in traditional medicine for their tonic properties.

Đỗ Tất Lợi is the first and only Vietnamese scientist to be honored with an honorary doctorate-science from the USSR. He was also given the Hồ Chí Minh Award in medicine and pharmacy. He published over 200 research works in Vietnamese, Russian, French, German, and Romanian including *Dược liệu và các vị thuốc Việt Nam* (Vietnamese Materia Medica and Medicaments), *Sổ tay dùng thuốc và chữa bệnh* (The Use of Medicines for Curing Disease), and *Thuốc Nam thường dùng* (Common Vietnamese Medicines).

Đỗ Tất Lợi's masterpiece remains *Những cây thuốc và vị thuốc Việt Nam* (Vietnamese Medicinal Plants and Medicaments). At 1257 pages, its eighth edition was published in 1999. It is the result of over 20 years' work. When the first edition appeared in 1962, Minister of Health Phạm Ngọc Thạch noted, "This is an excellent work, easy to read, and rich in content. It presents the author's personal experiments, the experiments of other Vietnamese people and people from other countries."

He includes in his work not only scientifically analyzed medicaments but also medicaments recognized as effective in clinical practice, whose effects have not yet been able to be studied scientifically. Thus, each medicament is presented in several aspects according to the conception of botany and botanical chemistry. The author also collected encyclopedic information. For example, concerning stag horn (*nhung hươu*), he mentions the rearing of stags in Việt Nam, in Siberia, and in China, the present state of the research, the drug Pantocrin created by Pavlenko, and popular recipes from stag horns.

In a 10,000-word article appearing in a journal of the Academy of Sciences of the USSR, L.L. Brekman, A.F.

Hammerman, I.V. Gruviski, and A.A. Yasenko Khmeleveski review the previous works of French authors on the Indochinese medicinal plants. They conclude, "Among the numerous works about tropical medicinal plants, none is equal to Đỗ Tất Lợi's from the point of view of accuracy, richness of details, and scientific value. Many plants which are presented there have appeared for the first time in the pharmacological literature."

ĐÀO VĂN TIẾN – A PIONEERING BIOLOGIST

In spite of the pouring rain, a large crowd of people gathered in the lecture hall of Việt Nam National University, Hà Nội on May 3, 1995 to pay their last respects to Professor Đào Văn Tiến. As Former Dean of the Faculty of Biology and Honorary President of the General Association of Biology of Việt Nam, Professor Tiến had attracted a large following. Over 100 delegations from research institutes, universities, and government offices came to the funeral, not counting students and colleagues, among whom were notable scientists in their eighties and nineties such as Nguyễn Xiển, Nguyễn Lân, Nguyễn Thạc Cát and Nguyễn Văn Chiển. People abroad also gave their condolences. Colin Groves, Professor at Australian National University, wrote a tribute to Đào Văn Tiến in the *Laboratory Primate Newsletter* (July 1995), stating that the role played by this well-mannered and mature man could not be overestimated in the history of Vietnamese science.

Born in 1920, Đào Văn Tiến came from a family of Confucian scholars in Nam Định. His father gave him a strict education, inculcating a sense of hard work and discipline from early childhood. When he attended primary school, he learned Vietnamese and French, but at home his father taught him Chinese characters through a textbook in three-word verses, which teaches Confucian ethics and, in particular, perseverance in studies.

In 1944, he graduated from Hà Nội Science University with a bachelor's degree in life science. After completing a graduate course in zoology, he published five research works in French in collaboration with Professor B. Noyer. At the age of twenty-five, he was invited to join the faculty of the same college where he had studied as lecturer.

Like other visionary intellectuals, he dreamed of elaborating a scientific lexicon in Vietnamese in order to build the base of Vietnamese science and teach science in Vietnamese at all levels up through university. Among the pioneers in this undertaking was Hoàng Xuân Hãn, a Paris-trained professor who published a lexicon, *Scientific Terms*, in 1942, covering the fields of mathematics, physics, chemistry, mechanics, and astronomy. While still a 22-year-old student, he told Professor Hoàng Xuân Hãn of his idea to add life sciences to this lexicon. The professor encouraged him in this project, lending him the French-Chinese dictionary compiled by Wang Yunyu and Taranzano to serve as a reference.

He devoted himself to his book for the next two years, rarely going to sleep before 1:00 a.m. To improve his knowledge of Chinese characters, he took private lessons at the home of a classical tutor. When he completed the manuscript of *Scientific Terms for Life Sciences* and sent it to the publisher, he immediately fell ill. In October 1945, two months after the August Revolution, his book was printed by the Union of Students for National Salvation.

In his preface to the book, Professor Hoàng Xuân Hãn wrote, "The author is one of the first group of young people to graduate from Hà Nội University in life sciences. The works of young people such as Đào Văn Tiến are worthy contributions of the young people whose duty is to unite to build an independent Việt Nam."

When the war of resistance broke out on 19 December 1946, Đào Văn Tiến was still affected by illness. However, he still left Hà Nội in the night for the guerrilla base of Việt Bắc. Soon after that, he enlisted in the People's Army. Dressed in uniform, he gave lectures at the Military School of Health and collaborated with Dr. Đỗ Xuân Hợp on a Vietnamese lexicon on anatomy. In 1951, following the mountain paths of the border zone, he arrived in Nanning, China, and was part of the group that built the Educational Quarter of the College of Fundamental Science and the Central Teacher Training College.

After the victory at Điện Biên Phủ and the liberation of Hà Nội, he returned to the capital. He was thirty-five years old then. He was appointed director of an important research program on the fauna, parasitic fauna, and entomofauna in North Việt Nam. After over twenty years of intensive research, he discovered ten new species of animals. He was first to give a description of two subspecies of leaf-eating langurs and found the first black gibbon in the Red River Delta and on the Chinese border in the northeast. His *Investigation of the Animals of Northern Việt Nam* and Van Penen's *Preliminary Determination of the Animal Species of Southern Việt Nam* are favorite books of specialists in his field.

During his career, he published over 100 articles in specialized international journals and kept up active correspondences with biologists from all over the world. This was a real achievement, considering how difficult it was to send letters between Việt Nam and many foreign countries at that time. As a scientist with an international reputation, he was invited to deliver lectures at universities in France and several other countries and was appointed a member of the Western Pacific Fishing Commission, the International Committee of the Study of Mammals, and an honorary member of the International Society for Animal Research.

Combining science and conscience, Professor Đào Văn Tiến always cared for the personal character of his students, following the principle of "not only teaching a trade, but also forming a person." After long years of reflection, he wrote *Applying the Scientific Method to Thought, Work, and Study*, published by Thanh niên (Youth) Publishing House, with the following dedication: "To young people, the hope of the country." As he recommended to students, "A career that you build for yourself will always be more solid than one you owe to the help of others."

A steady and thoughtful man, he listened much, spoke little, and disliked long speeches. Both his specialized books and his articles on popular science are precise, rich in content and full of original ideas. His popular three-volume series, *Hỏi đáp về động vật* (Questions and Answers about Animals), became a favorite of young people during the American war in Việt Nam. Young women in the Volunteer Youth used it to identify unknown animals along the Hồ Chí Minh trail. Several soldiers wrote him to settle a question that they had argued about on the trail, namely whether the correct name of one species of boa constrictor has "skin like the mesh of a hammock" (*trăn mắt võng*) or "a body shaped like a hanging hammock (*trăn mắc võng*). The professor answered from Hà Nội that the first name is the right one.

After reading Professor Đào Văn Tiến's works on animals, one student from the Dao ethnic group in Cao Bằng Province wrote in the *Thiếu niên tiền phong* (Young Pioneers) magazine that he felt pity for wild birds and decided not to capture them any longer. Also, the well-known writer Tô Hoài once declared, "All the experience of animals that appears in my writings has been drawn from the work of Đào Văn Tiến."

In 1996, Professor Đào Văn Tiến was granted a posthumous Hồ Chí Minh Award, the highest award of the Vietnamese Government, for one of his final works, *Tập hợp các công trình điều tra cơ bản về động vật học ở Việt Nam, 1957-1980* (A Compilation of Basic Surveys on Zoology in Việt Nam, 1957-1980). It is no wonder that Đào Văn Tiến continues to enjoy popularity even after his death.

LÊ TÂM – INVENTOR OF THE ANTI-TANK SS GUN

In 1945, a young intellectual group who wholeheartedly supported Hồ Chí Minh's Government was founded in Paris. They included Phạm Quang Lê (also known as Trần Đại Nghĩa), Trần Hữu Tước, Nguyễn Khắc Viện, Phạm Huy Thông, Lê Văn Thiêm, Trần Đức Thảo, Hoàng Xuân Nhị, Lê Tâm, and some others.

After the Fontainebleau Conference failed to reach any agreement on Indochina's independence, Lê Tâm was sent to Italy to gain experiences in water management and to secretly purchase weapons.

In December 1946, he returned to Sài Gòn (present-day Hồ Chí Minh City) and was assigned to take charge of the Southern Military Weapons. He succeeded in inventing the SS gun (also known as Sác Forest gun), which is used against French tanks. He won the Hồ Chí Minh Award for this invention.

First-rank High-school Graduate in Indochina

Lê Tâm (born Nguyễn Hy Hiền) was born in 1921 in a traditional Confucius family in Niêm Phò Village, a famous village in Quảng Điền District, Thừa Thiên-Huế Province. His father, Nguyễn Văn Mại, was one of the five most knowledgeable scholars in Việt Nam at that time.

He gained a lot of excellent achievements in study. In 1939, he was the first-rank high-school graduate in Indochina. Then, he received a scholarship from the Như Tây Overseas Study Society, a study promotion organization in Southern Việt Nam, which offered only one scholarship per year to study in France.

He left Sài Gòn for France on a ship with Lê Văn Thiêm. They met and talked to each other every day. They became close friends. In September 1939, the Nazis entered Paris; therefore, the practice of sending Indochinese students to study in France was discontinued for many years.

After finishing the pre-university course at Lycée Saint-Louis, he passed the entrance exam into École Nationale des Ponts et Chaussées (National University of Civil Engineering), a big university in France. A few years earlier, Phạm Quang Lễ had studied there. Every day, he had lunch with Lê Văn Thiêm and Trần Đức Thảo in a canteen in the Latin town, a few hundred meters away from École Normale Supérieure (literally translated Paris University of Education). (Lê Văn Thiêm and Trần Đức Thảo were studying at this university.) A group of young Vietnamese intellectuals in support of the Vietnamese Revolution was established in Paris.

In 1946, a delegation of the Democratic Republic of Việt Nam led by Mr. Phạm Văn Đồng was sent to France to attend the Fontainebleau Conference near Paris. President Hồ Chí Minh was the leader, but he did not join the delegation directly.

Phạm Huy Thông, Lê Văn Thiêm, and Lê Tâm volunteered to assist the delegation by collecting documents from archives and libraries in France to provide evidence against the French Government.

Two Special Tasks and One Ring

Lê Tâm told me about his strange experiences in 1946.

"In June 1946, Phạm Huy Thông, Chairman of the Association for Vietnamese Residents in France which was assisting the Vietnamese delegation at the Fontainebleau Conference, said to me, 'Due to France's obstinacy, the conference seems to fail to reach any agreement. We have nothing to do. Thus, Mr. Đồng wants to assign you a special task. You will go to Italy to learn about water management. In his opinion, Italy has many experiences in the Po River water management. This river and our Red River have many features in common. The experiences that you learn there will be very helpful for our government....'

At that time, France and Italy are in dispute over their border. It has nearly escalated into a full-scale war after the World War II.

I accepted the task. 'But for how long?' I asked.

Thông answered, 'As Mr. Đồng said, it is about four or five months because our delegation might have to return home at the end of the year. A provisional agreement won't satisfy us. You are supposed to be in Paris by then and come back with the delegation.'

I asked him about the cost for the trip to Italy.

He gave me a diamond-covered gold ring and said, 'Mr. Đồng asked the man who takes charge of our financial issues to give you enough money for the trip. Unfortunately, there isn't any cash left but some gold and this diamond ring at present. They were donated at the Gold Festival Week in Hà Nội. Exchange it (the ring) for cash to spend in Italy. Return to Paris when you run out of money.'

I was very surprised by this form of payment, but I still took it. I had just graduated from university and hadn't found any full-time job yet. I had earned just a little money by tutoring several French students for their entrance exams. Moreover, I had just been imprisoned near Paris for distributing leaflets against General Philippe Leclerc's decision to send French forces onto the Sài Gòn Harbor. Trần Đức Thảo had been seized two weeks earlier for the same reason. Thanks to the French communists and the left-wing party's help, Thảo and I were released, but I had no money.

Before leaving Paris, Mr. Tạ Quang Bửu, Vietnamese Minister of Defense and a member of the Vietnamese delegation then, said to me, 'This is your second task. Try to contact the anti-fascist guerilla groups there and ask them whether they can sell us their unused weapons.'

He assigned this task to me without mentioning money to buy those weapons."

Penniless in a Big City

"With the ring on my finger and a little savings, I bought a train ticket from Paris to Rome. The train went through two of what were then the world's longest tunnels at the foot of the Alps. In my bag, there were some letters from

Bửu Hội, a member of the Vietnamese delegation, to ask his intellectual friends to help me to meet the Prime Minister of Italy at that time.

When I arrived in Rome, a friend of his, a famous professor at Rome University, took me to the Palace of Prime Minister on his bike. In 1946, life after the World War II was hard, so it was normal for professors to get to work by bike.

The Prime Minister was hosting a conference, but he stopped for a while to meet me. He said briefly, 'Italy is in conflict with France over the border and Việt Nam is struggling against France for independence. Therefore, Italy and Việt Nam are friends.' He considered me the first Vietnamese 'ambassador' to Italy.

He promised to introduce me to the Po River water management officials. I was very delighted to hear this. However, I still had to deal with living in an unfamiliar city by myself. The ring became the one and only solution."

Learning about Water Resources Management and Finding Ways to Buy Weapons

"I was still looking for accommodation in Rome when I come across a priest who looked Vietnamese in Vatican. I got up my courage to speak to him. After finding out that he knew the priest Luận in Paris (a member of a Catholic group against France), I asked for his help in selling the ring. He took me to a jewelry shop. The shop owner checked it and said to us that he needed a value assessment paper from Lloyd in London. Then he would decide whether or not he would buy it. He suggested sending the ring to London. I had no choice but to accept.

While waiting for Lloyd's assessment, everything, including accommodation, food, transportation, and courses in Italian, was paid for by my own small amount of money.

The priest helped me find accommodation at his friend's house. I didn't have to pay the rent immediately, but I still had to find my own food. The Western culture is very different from my own. Therefore, I had to go out for meals. Breakfast with a slice of bread, lunch with two slices of bread and some sauce, and dinner with a bowl of noodles without meat were my meals at that time.

Ten days passed slowly. While I ran out of money, the priest suddenly appeared and cheerfully said, 'Lloyd has answered. The ring is worth 18,000 lire.'

No words could describe my mood then. I trembled with happiness. We went to the jewelry shop together. The shop owner told me that it was just the value assessment. Whether I sold it or not was up to me. Although, to be honest, I was very pleased, I answered coldly, 'The price seems to be rather low, but I take into account your sending it to London. I accept it.' I took 18,000 lire. The priest told me that, with such an amount of money, I could live comfortably in Italy for four months.

After a short time, I received an announcement from the Italian Prime Minister's office that I would meet the Po River's drainage inspector. This river flows 562 kilometers from the Alps through a delta and into the Adriatic Sea. Thanks to him, I went to many places in Northern Italy to gain experiences and learn from many engineers.

The professor who took me on his bike turned out to be the man who had worked secretly for an anti-fascist group. He was delighted to introduce me to guerilla groups which had old weapons. They said they would give away the weapons for free if we could afford the shipping. I reported this matter to the defense's top leaders.

When Lê Tâm went back to Paris, President Hồ Chí Minh and the delegation had already returned to Việt Nam. Thus, he came back to Việt Nam alone."

Taking Charge of the Southern Military Weapons

On the evening of 19 December 1946, the ship that Lê Tâm was on docked at the Singapore Harbor. At that time, the Việt Nam War against France broke out. On 23 December, the ship docked at the Sài Gòn Harbor. A month later, Lê Tâm worked with the staff of the Southern High Command in charge of the Southern Military Weapons. In 1949, he became one of the seven officers who were promoted to colonels. (Phạm Quang Lễ was promoted to Major General at the same time.)

In the Southern battlefield, the military workshops were replaced frequently, sometimes in Đồng Tháp Mười or sometimes in the mangrove forest of Bà Rịa-Vũng Tàu. This forest is home to *sác* trees that can stand high salinity, so it is also known as Sác forest. When France attacked, the Vietnamese buried the weapons that they were working on. When the enemy left, they would dig them up. There were many difficulties and hardships, but they were still successful in building big military workshops such as the workshop of the Southern High Command with nearly one thousand staff members and workers and that of Military Zone 8 with three hundred workers. The forces of the Southern Military Weapons increased to eight thousand members.

Invention of Sác Forest Guns Against French Tanks

Engineer Lê Tâm envisioned a gun which was capable of destroying a cannon but also light enough (about 5-10 kilograms) for soldiers to hold on their shoulders. To make it, he had to produce heat bullets, which had just appeared in the World War II. On explosion, a heat bullet could force energy into a minimum area. The heat could reach 3000°C with an extremely high pressure. With such heat and pressure conditions, it would be much easier to penetrate a tank shell. Thus, all the enemies in the tank would be shot. A heat bullet was very big, only the tail fit in the barrel. Lê Tâm called the gun the SS gun (or Sác Forest gun).

In the beginning of the war, French troops often sat comfortably in tanks and warships when attacking. Rifles and grenades did little damage to the exteriors. However, the SS gun was able to penetrate their exteriors. In small canals in the Southern areas with a width of eighty meters, a soldier could take a SS gun to shoot at the ships. The distance of forty meters was enough for bullets to make an impact. Through this, many ships were damaged and sank. Lê Tâm said, "We had only used the SS gun for some battles, but French troops didn't swagger any more."

Destroying stone and building a road to Điện Biên Phủ

In 1952, Lê Tâm was ordered by the government to walk continuously from Southern Việt Nam to Việt Bắc. To receive weapons from China, it was necessary to build a new highway which was eight meters wide. During French colonization, highways in Indochina were normally 6.5 meters. Actually, the new highway ran from Mục Nam Quan and across Đồng

Đăng (Lạng Sơn Province), Bắc Sơn, and later Đình Cả (Thái Nguyên Province). It was totally 150 kilometers long and was completed after six months of construction. In Indochina at that time, there were only two mine specialists, Lê Tâm (graduated in France) and Nguyễn Văn Thái (graduated in Indochina). Therefore, the government ordered Lê Tâm to journey from the South to Việt Bắc.

Peace was gained in Northern Việt Nam. Lê Tâm returned to Hà Nội and led a group to repair the railway networks of Hà Nội - Mục Nam Quan and Hà Nội - Vinh. He later became Dean of Civil Engineering at Polytechnic University, Head of Metrology Department, and Member of State Science and Technique Committee of Việt Nam (present-day Ministry of Science and Technology).

In 1986, he and his collaborators in the Sác forest during the anti-French resistance war were given the Hồ Chí Minh Award by the Vietnamese Government for the invention of the SS gun.

ĐẶNG HỒNG VÂN AND THE NATIONAL PHARMACY

Đặng Hồng Vân was born into the family of a civil servant in Vinh (Nghệ An Province) during the era of French colonization. In 1935, when she was ten years old, she and her brothers and sisters followed her parents to Hà Nội. Her three brothers went to public schools while she and her sister attended a private school. In the private schools discipline was less strict, so the girls could be absent when they were needed to do housework.

Following her father's decision, her sister got married at the age of sixteen. She, one year older, tried to delay her marriage. "Dad, permit me at least to finish my baccalaureate," she pleaded. Once she had graduated from high school, she studied pharmacy.

When the war between France and Việt Nam broke out on 19 December 1946, her family had to leave the capital city for Nghệ An Province.

Hearing the thunder of guns while taking refuge in Mễ Trì Village near Hà Nội, she decided her life's direction. Like many of her friends, she enrolled in the army medical service in Thái Nguyên, an old bulwark of the revolution.

The French attacked Thái Nguyên. She moved further north and joined the Bắc Cạn Civilian-Military Hospital under Dr. Dương Văn Ân, a Southern Parliamentary member. Her daily task was to distribute medicines and produce malaria quinoserum. Her jet-black hair, rosy complexion, and young smile were attractive to many students. Among them was a southern lawyer, Trần Công Tường. Never having been in love before, she began dating him in the small town inhabited by the Nùng ethnic minority.

But the 1947 French strategic offensive separated the lovers. She left Bắc Cạn for Thái Nguyên where she worked at a laboratory producing medical supplies for the Việt Bắc Special Zone. She, together with her former schoolmate Đoàn Hữu Sử, had to train dozens of pharmacy assistants.

Some time later, she received a letter from her father, "I am very anxious about you, and I can't sleep at night. Your brothers and your sister have come back to Nghệ An to teach. Our province, part of the free zone, is untouched by the war. I'm told that French troops have parachuted on Bắc Cạn and this has increased my worries because you are a girl, not a boy. If you fall in the hands of French troops, what will come of it? I dare not think any further. If you have pity on me, come and join us."

She felt trapped in a moral dilemma between her sentiments towards her father and her commitment to the revolution. She finally chose the second.

There was another dilemma: she had accepted Trần Công Tường's marriage proposal. The wedding had to take place quickly because of the war situation. Her father was not informed since enemy encirclement had destroyed the postal network. According to the Vietnamese tradition, a child cannot get married without paternal consent. But her father had never wanted his daughter to marry someone that was not a local, especially not a southerner.

After considering the matter, she wrote to her father to defend her actions. She regretted displeasing her father, but she couldn't do otherwise. She remained in the resistance movement after her marriage.

At the end of 1948, she gave birth to a boy. One month after her delivery, she decided that she wanted to study at the Pharmaceutical Department of the Institute for Military Medicine where the well-known pharmacist Dr. Trương Công Quyền was teaching. Her husband tried to keep her by his side. He said, "I'm going to find you work as a secretary for the government. Then you'll be able to care for the baby properly. You can go on studying after the war."

Not wanting to be too distant from her classmates, she carried her baby on her back when travelling on foot to Chợ Mới. There she taught at the Military Pharmacy Secondary School. In 1951 she had another child, a girl. She persevered in her task, always living far from her husband. In 1952, she crossed 200 kilometers of mountain trails to attend a refresher course on pharmaceutical medicine with her daughter on her back and her son at her side.

The University of Medicine and Pharmacy and Chiêm Hóa Hospital were composed of bamboo stilt-houses on a hill. At the foot of the hill flowed the Quẵng Stream. People ferried across the stream to reach a highland village which sheltered students and professors. Professor Hồ Đắc Di was director of the college, surgeon Tôn Thất Tùng was in charge of the hospital, and Health Minister Hoàng Tích Trí was in charge of microbiology. The proximity of the forest allowed for an easy escape in case of enemy attacks. In order not to bother others, she built herself a small stilt-cottage. It is impossible to list all the difficulties encountered by this student and young mother. She often had to ask a woman evacuated from Vĩnh Yên to care for her children, especially her one-year-old girl suffering froṃ malaria. At midnight, she had to come home alone on the ferry. Everyone was under-nourished. There was almost no meat, only wild vegetables for a main dish.

The eleven-student class was run by four physicians and one chemist. Despite shortages of all kinds, it functioned according to the required standard. It made an effective contribution to the resistance by producing medicines from local medical ingredients such as chili, Passiflora hispida DC., garlic, Amomum xanthioides Wall., Datura meṇtal L., Strychnos nux vomica Linn., and cinnamon. In particular, thanks to mushrooms brought back from Japan by Dr. Đặng Văn Ngữ, the college could produce penicillin which was precious during wartime.

In 1953, she graduated as a first-class pharmacist with an excellent record. She stayed and worked at the hospital under Tôn Thất Tùng. She had no opportunity to meet her husband who worked near her house but was separated by a mountain and a large river.

After the Điện Biên Phủ victory, she came back to liberated Hà Nội to teach at the Pharmaceutical University. She was worried that her knowledge was limited due to her irregular training during the nine years of resistance. In her field, Vietnamese books were not available, and French and

English books were out of date. But there was no money to purchase new editions. She decided to learn Russian so that she could read pharmaceutical works from Moscow. She was no longer young, but with great effort she succeeded in reading Romanian and Bulgarian specialized journnals, two languages similar to French. She also attended an in-service course on chemistry at Hà Nội University.

In 1964, American aircrafts began bombing North Việt Nam. The Pharmaceutical University moved to Sơn Động in Bắc Giang province. At that time, she was being treated in the hospital. She now had three children, the eldest daughter being fourteen years old. She could have asked to be moved to a research institute in the capital city but she could not leave her colleagues and students or her posts as Leader of Discipline. Once more, she chose the difficult path and moved back to the mountain. She did so with firm determination to inspire female students. After the 1945 Revolution, opportunities for girls in education increased, and there were sometimes more females than males in classes, particularly in medicine, pharmacy, and pedagogy.

In the anti-French resistance war (1946-1954), the civilian population was evacuated from cities with rudimentary vehicles, often on foot. At the beginning of the American air war, the civil service had a number of trucks at their disposal. This was why she, her colleagues, and students at he Pharmaceutical University were easily brought to a new destination in the northern mountains. The students of the college were divided into small groups. While studying, they worked in hospitals, pharmaceutical production factories, and field pharmacies. During the years of intense bombing, students made contributions to the war effort by preparing serums and medicines. Professors who did not have regular laboratories created "portable labo-trunks" and bicycled long distances to meet their students. She lived with her younger daughter in a highland village. Here the child studied at a primary school with Tày ethnic minority children.

After thirty years of resistance against the French and the Americans, peace was finally reestablished in Việt Nam in 1975. Being close to fifty, hardened by innumerable trials, and experienced as a teacher and researcher, she had proven herself qualified for large-scale scientific projects. The Ministry of Health gave her the task of researching the preparation of medicines from Vietnamese medical ingredients. It is the official policy of the Ministry of Health to combine Western and traditional medicine.

Traditional Vietnamese medicine is made up of two components: southern medicine, which is wholly Vietnamese and composed of popular indigenous recipes, and northern medicine originating from China, which is more scientific. For thousands of years, this Sino-Vietnamese medical system provided healthcare for the majority of people, using a collection of conventional concepts for an empirical and effective treatment of patients. The scientific preparation of traditional medicines in the form of pills, solutions, and ointments requires exploitation of medical ingredients from the local fauna and flora.

Research on local medical ingredients never stopped attracting her. She immediately began to work, continuously widening her field of exploration and looking all over the north and the south of the country.

Her first great success was in the use of seven species of sea snakes (*đẻn*), three in the north, and four in the south, for the production of a cure for rheumatism and for tonics. Every year, central and regional pharmacies prepare hundreds of liters of traditional tonic alcohol based on snakes and even this cannot meet the customers' needs. While raising land snakes had been an insignificant activity, the culinary demand for snakes for gastronomy made them extinct. It is not surprising that medicines prepared from snakes are expensive.

She had thought of looking at sea snakes since 1971. This idea had never been conceived by any other researcher. She studied in fishermen's villages up and down the coast of Việt Nam. Staying in a fishing cooperative in Cửa Lò (Nghệ An Province), she reached precise conclusions. Fishermen in the regions caught between twenty and thirty tons of sea snakes every fishing season. These snakes were usually thrown back into the sea because people rarely ate them. Fishermen were ready to sell them as the same price as fish, which resulted in the cost of medicines prepared from sea snakes being much lower than medicine from land snakes. This was a good area to explore. But were the curative effects of sea snakes the same as land snakes? Numerous laboratory tests on mice and human subjects carried out under her leadership gave an affirmative answer. Finally, alcohol prepared from sea snakes gained the public's confidence. The problem of medical ingredients in this field was solved.

But the issue of sea snakes still interested her. A two-week holiday in Nha Trang gave her an opportunity for in-depth study. The Oceanographic Institute, founded in 1932 by French scientists, had collected some 600,000 samples of marine fauna up to 1977. She reserved all her time observing the living beings and the exploration of the institute's vast library. She was particularly interested in the works of Bourret, Shuntov, and Barme on Indochinese sea snakes. But she did not find any reference to the use of snakes as medical materials. Therefore, she was the pioneer in this field.

There were a number of issues on this subject. What were the main species of sea snakes in Vietnamese waters? What was the numerical potential? Did the flesh, skin, and bones of these species have the same properties as land snakes used in the preparation of traditional medical alcoholic drinks? How effective were their bile and their venom?

As medical alcoholic drinks prepared from snakes did not suit a number of patients, especially those who could not drink alcohol, this medical material was to be prepared in the form of jelly (*cao*) or pills. The experiment of snake pills on rheumatic patients at the Friendship Hospital proved the efficacy of the medicine. Her team produced a traditional drug against coughing prepared with snake venom and dry tangerine rinds in the form of alcohol and syrup. Cobra venom, which is expensive but had been rejected by laboratories, was now collected for preparing anti-rheumatic ointment and analgesics to be injected in cancer patients.

Among her list of pharmaceutical breakthroughs, there are other achievements. Ramasol is a Vietnamese product prepared from pennywort (*rau má*). Exported to England, the Netherlands, Italy, and Canada, it is now considered effective in healing wounds and quickening the process of recovery of tissue and organ lesions, which is very useful for surgical operations.

The use of *nhàu* plant (morinda citrifolia) has been another contribution of Đặng Hồng Vân. Infusions of *nhàu* roots have been a popular prescription for treating backache, rheumatism, dizziness, and high blood pressure. *Nhàu* fruit taken with salt can also regularize menstruation and treat uterine hemorrhage.

She thought that *nhàu*, being rich in organic selenium, might help treat cardiovascular diseases, stimulate the immune system, fight against cancer, muscular atrophy, and delay the process of senility. This sensitive plant had been discovered by chemist Đàm Trung Bảo.

The idea of using *nhàu* came to her when she stayed in Nha Trang. Out for a walk, she saw the plant in a garden for the first time. The owner was glad to give her sample. She filled her suitcase with its leaves and stems when she returned to Hà Nội.

She exerted her influence thanks to the network of her students. During the American air war against North Việt Nam, many of them secretly went to the south to serve the resistance. When peace was reestablished, they became leading cadres of pharmaceutical production in the south. As a result, all products made by her have gone south.

Her pharmaceutical contributions have also been welcomed abroad. At a session held in December 1989, the French National Academy of Pharmacy elected Đặng Hồng Vân as a correspondent member. This honor also constitutes a reflection on Vietnamese women in general. Up to the end of the nineteenth century, no women could attend the triennial civil service exams. Among 2,874 doctoral laureates, there was not a single woman. During the French colonial regime, the only woman doctor was Hoàng Thị Nga, who had a doctorate in physics. Even today, women professors are not numerous.

Unfortunately, not longer after her election to the academy, Đặng Hồng Vân died following a road accident leaving a great hole within the ranks of Vietnamese intellectuals.

CHAPTER TWO

THE SECOND GENERATION OF MODERN VIETNAMESE SCIENTISTS

ĐẶNG VĂN CHUNG WITH THE DIFFICULT ROAD

"North or South? Should I stay in the North or leave for the South?" Dr. Đặng Văn Chung racked his brain to find the solution to this terrible dilemma.

It was in August 1954. After the fall of Điện Biên Phủ in May, the Geneva Agreement, signed in July, ended the war of reconquest fought in vain by the French Expeditionary Corps for nine years. Waiting for the general election to reunify the country that was supposed to take place two years later – the election later sabotaged by the Americans – Việt Nam was divided into two zones at the seventeenth parallel. The North returned to the resistance government of Hồ Chí Minh, the South to a pro-French administration, or rather Franco-American, soon to become exclusively American with Ngô Đình Diệm. A cooling-off period of 300 days was planned to allow for the regrouping of both sides' forces. The revolutionary government would not take control of Hà Nội until 10 October.

While they were waiting, the pro-French administration remaining in the North put pressure on all its officials to move to the South. Đặng Văn Chung, an eminent physician, was one of these officials. After a long reflection and discussion with his wife, he finally decided to stay in the North. A letter sent by his former colleague at the Hà Nội Medical College, Tôn Thất Tùng, who was vice-minister of health in the revolutionary government, reinforced Chung's determination. He would stay in the North, running the risk of failure. Would the resistance accept him after he had left the Medical College in the liberated zone to return to the French-occupied capital and then proceeded to France to complete his medical degree?

The government led by Hồ Chí Minh, which had convicted Chung in absentia for abandoning his post, welcomed "the prodigal son" with open arms. This is because the doctor had chosen to give up a life of luxury to place himself at the service of his country which was still facing numerous external and internal difficulties.

Born in the southern town of Sa Đéc in 1910, Đặng Văn Chung had studied for his baccalaureate at the Lycée Chasseloup-Laubat in Sài Gòn. After completing a medical degree at the Medical College in Hà Nội, he continued his studies in France. When he returned to Việt Nam, he soon acquired a reputation as a general practitioner. But it was only after he rejoined the revolutionary government that he was able to use the full measure of his professional competence and his qualities as doctor in the service of the people.

As a doctor at the most important clinic in the North, a researcher, and professor at the Hà Nội Medical College, he had an extremely busy schedule. At Bạch Mai Hospital, he had consulting hours every morning from Monday (cardiovascular patients) to Tuesday (respiratory ailments), Wednesday (endocrinology), Thursday (blood and bone ailments), Friday (urology), and Saturday (digestion).

During this period, there was a general lack of specialists in internal medicine. He astounded the specialists by the depth of his knowledge in each of the specializations.

Even today, after his death, there is no shortage of anecdotes about his medical skill. One morning, the professor was making his rounds in Building C of Bạch Mai Hospital. The doctor on duty invited him to examine a "special case." The patient was seated with his legs crossed on his bed, which was full of bananas, candy, and packages of biscuits.

"Why don't you put all this stuff in the small cabinet there at the head of your bed?" asked the professor, intrigued.

"Because I don't have enough time, Doctor," the patient answered.

"Not enough time for what? I don't understand."

"When I'm starving and I can't get something sweet in my stomach, I go unconscious."

The doctor on duty stepped in, "He has a very strange disease, Professor. He's hungry all the time, and if he doesn't get food, he goes into convulsions like an epileptic. In his neighborhood they call him 'Mr. Hunger.' Sometimes he starts destroying everything around him looking for food. His family brought him to the mental ward."

After examining all the patient's clinical and laboratory records, Dr. Đặng Văn Chung concluded, "I don't think this is a mental case. It may be that this patient has caught a disease that no one has seen before in North Việt Nam or maybe anywhere else in Indochina. We have to check it out further."

He remembered having read sometimes during his medical studies about a type of hypoglycemia caused by a tumor of the pancreas. But no case of this malady was recorded in the medical archives of Indochina. X-rays of the patient could not reveal any tumor larger than the size of a grain of rice.

He gave the patient sugar water to drink as a temporary measure. But no cure was possible without surgery. He consulted his colleague Tôn Thất Tùng, a world-renowned surgeon, who agreed to his request. But Dr. Tôn Thất Tùng completed the operation without finding any pancreatic tumor. Was there a diagnostic error? No, Dr. Đặng Văn Chung was sure he could not have been wrong. Dr. Tôn Thất Tùng examined the tip of the pancreas one more time and finally discovered a small tumor that he removed with the aid of a magnifying glass. And the patient was cured of his pathological hunger.

Besides this hypoglycemic tumor, Dr. Đặng Văn Chung uncovered a number of other almost unknown diseases in Việt Nam at that time: pheochromocytomes, Seckel's disease, and Takoyashu's disease. Numerous foreign doctors who worked in Việt Nam have attested to his professional capabilities. Among them included Russian academician Ado, French Professor Francois Cannes, and Hungarian Professor Peter Varga.

Immediately after the liberation of Hà Nội in 1954, he began editing several university-level textbooks such as *Bệnh học nội khoa* (Internal Pathology) and *Điều trị học* (Therapeutics), both considered classics in the field. As the supervisor of internal medicine at the Hà Nội Medical College he wrote *Symptomatology of Internal Medicine*, a work of capital importance for teaching. In collaboration with pharmacologist Trương Công Quyền, he also edited *Tra cứu y-dược* (Compendium of Pharmaceuticals). He also did not neglect popular medicine, publishing books on the prevention of heart disease and the practice of good general health.

Almost all Vietnamese practitioners of internal medicine are thus directly or indirectly disciplines of Đặng Văn Chung. In recognition of this fact, he was elected president of the Internal Medicine Association of Việt Nam.

During the American bombing the professor remained at Bạch Mai Hospital to care for the wounded and for emergency patients in the basement. Even though he was more than sixty year old, he continued to work in his office on the second floor rather than move to the underground shelter. One day, while he was sitting on a stool between two hospital beds, a bomb exploded only ten meters away, throwing him onto the floor and decapitating a young doctor who was standing near the window. Half of the building collapsed. He was able to return to his office to collect the papers of his manuscripts that had been scattered all over the floor.

On the night of December 22, 1972, American B-52 Super fortresses dropped bombs aiming to destroy Bạch Mai Hospital. Building A9 – the main building overlooking the hospital gate – was bombed to pieces. There were numerous casualties. When he looked back at these tragic days, he did so with emotion mixed with pride.

One day in autumn 1978, I had the pleasure of visiting him at his home. In his retirement, he lived at the end of an alley lined with moss-covered walls. The professor talked emotionally about President Hồ Chí Minh's visit to his home sixteen years before. He then showed me a photo.

"Here he is seated together with the members of my family. That was in the Lunar New Year's festival in 1962. He came to wish us a happy new year. Look, he really does look like an affectionate uncle. Can you see this little girl with short hair who's holding her hand out shyly for a piece of candy? She was born one year after the liberation of the capital. That was twenty-three years ago. Just this morning, she received her medical diploma."

At that moment, the girl herself, Đặng Kim Châu, came in. She was now a sparkling young woman with long shining hair and thin figure. Drinking tea with us, she recounted her investigations of the health conditions of ethnic minority people in a mountainous area near the border. Her father followed her words with the attention of a colleague.

I thought of the novel *The Road to Calvary* by Russian writer Alexei Nikolaevich Tolstoi (1883-1945) concerning intellectuals who regained their peace of mind after choosing to follow a revolutionary path. In following this difficult road, Đặng Văn Chung's soul came to rest.

NGUYỄN VĂN TRƯƠNG – A WATER AND FORESTRY ENGINEER IN A PRODUCTIVE RETIREMENT

From Ba Vì Mountain to the Old-Growth Forest of Mai Châu

After he retired in 1990, Nguyễn Văn Trương created the Institute of Ecological Economics (IEE), the first private scientific institute in Việt Nam. This meant that the director and managers had to take care of themselves and provide for the multiple expenses of the institute: only 170 million *đồng* (US $11,000) was given per year for eleven salaried staff and administrative expenses, not to mention professional equipment and services (computers, cars, and laboratories). The institute had to rely on the almost charitable assistance of nineteen professors, science Ph.D.'s, and retired officials, who accepted only a nominal payment for their contributions. For the past thirteen years, the state has not given the institute any money.

"How have you been able to keep so many projects going?" I asked.

"We rely on the international organizations which support two or three projects of ours every year," he answered, "about US $30,000 a year. Thirty percent of this sum is enough to keep the institute going, with administrative charges reduced to minimum. The rest, 70%, is entirely devoted to the projects. In a good year, with receipts mounting to US $40,000-60,000, we are able to pay for computers, motorbikes, and cars to go to the countryside or to distant mountainous regions."

"But how can you obtain donations in cash?"

"Of course, it is thanks to the prestige of the members of our institute, many of whom are internationally-known. In order that a request may be funded by a foreign humanitarian or scientific organization, we must answer twenty or thirty questions in a convincing manner. I can make a short list of our donors: the Catholic Committee against Famine and for Development (CCFD), the World Conservation Union (IUCN), and the International Federation of Organic Agriculture Movements (IFOAM), among others."

In order to understand the contributions of IEE, I visited the Hợp Nhất cooperative at the foot of Ba Vì Mountain, 65 kilometers from central Hà Nội. This is a cooperative of the Dao ethnic minority. Mr. Dương Trung Tâm, chairman of the cooperative, told me, "We owe everything to Professor Nguyễn Văn Trương. He sent us a forestry engineer, Nguyễn Đình Cầm, who directed all the work we had to do: rice-planting, gardening and orchard planting on terraces, digging wells, planting yams, sugar cane, fish breeding, and beekeeping."

The story of this retired engineer who traveled from central Hà Nội twenty days a month on an old motorcycle to stay in the countryside moved me so deeply that I stealthily wiped a few tears away when I hugged him at our first meeting. He wanted to help the Dao, whose lifestyle consisted of nomadic slash-and-burn agriculture, move to a stationary lifestyle.

I learned about another example of IEE's contributions when I visited the Thái people of Mai Châu (Hòa Bình Province), accompanied by the retired economics professor Nguyễn Pháp and the IUCN coordinators, Mr. and Mrs. Jacques Lecup.

"I would like to help donors see the poverty of the mountain people through their own eyes, as well as the local officials' ability to make good use of their aid," said Professor Nguyễn Pháp.

In Chiềng Châu Commune, 87% of the 2.132 hectares of land area is forested. Professor Nguyễn Pháp also invited the guests to visit the old-growth forest. The Lecups had worked for three years in Nepal, at the foot of the Himalayas. These passionate mountain climbers were ready to support a forest preservation project.

From a Kim Liên Schoolboy to a Ph.D. in Dresden

Nguyễn Văn Trương was born in 1922 in the same village as Hồ Chí Minh, Kim Liên (Nghệ An Province). Nghệ An has long been famous as the homeland of matchless personalities including revolutionary patriots and other people of marked individuality and strong senses of purpose. His father, a poor schoolmaster, did his utmost to help his son continue his studies. Admitted to Khải Định Lyceum in Huế, he studied with the prominent mathematics teacher Nguyễn Thúc Hào, which gave him a solid base for his future studies in forestry. After finishing his secondary education, he went to Hà Nội to continue his study of higher mathematics. He was admitted to the Brévié College of Agriculture and Forestry after taking a competitive examination. He chose to specialize in forestry, as the twice-larger scholarship allowed him to send a stipend to his family in the countryside. As a young water and forestry engineer, he was then appointed by the colonial administration to manage a forest reserve in Hà Tĩnh Province and then another in Nghệ An.

The August 1945 Revolution put an end to the eighty years of French colonization. The resistance war began. Throughout this period, he taught in Tân Dân, a private secondary school. The teaching was carried out in unimaginable conditions of hardship. Asked about these unforgettable days, he said, "The communal house served as a classroom. The tables were made of bamboo trunks lashed together. Courses took place at night in the pale light of oil lamps. As no paint was available, blackboards were wood planks covered with soot. The teachers, coming from all parts of the country, dictated their lessons to students because there were no printed books. Copy-books were made from bamboo leaves or from used paper that had been soaked in lime water to wash out the ink. Students studied very hard and when they weren't studying they did to farm work. Many of them eventually became high-ranking military officers, journalists, physicists, and doctors."

When peace was restored in 1954, he was called to Hà Nội and appointed Director of Studies at the Hanoi Agriculture and Forestry College. He took the time to learn Russian and German on his own, on the basis that knowledge of these two languages, in addition to French and English, was indispensable to researchers. In 1958, after being appointed to the Ministry of Forestry, he had the chance to practice German and Russian with specialists. From 1964-1970, while working in the Forest Survey Department of the Institute of Forestry, he collected documents that would serve as the basis for his future doctoral thesis. Here, frequent contact with Chinese experts gave him the opportunity to learn modern Chinese, based on the knowledge of Chinese ideograms he had acquired during high school in the city of Vinh.

Foreign languages served as an instrument for him to explore his specialization, bioforestry, in more depth. Reading an article in French, "Revue forestière," on the Bitterlich method of conducting a quick inventory of the quantity of wood in a determined space of the forest led him to biometrics. He worked hard for several years to master this discipline, thanks to his solid mathematical knowledge. For the subsequent two years, he carried out investigations of all kinds of forests (natural and old-growth forests, as well as secondary growth forests) together with young mathematics colleagues. This research resulted in the development of a new method of statistical investigation of mixed forests. This method had many important practical applications. Before him, people only used the qualitative method with the distribution function of probability.

This problem was the subject of his 1978 habil doctoral dissertation at the Technical University of Dresden (East Germany), "Research on the Tri-dimensional Structure of Tropical Forests in Northern Việt Nam." Two years earlier, he had submitted his Ph.D. thesis titled "Mathematical Applications of Research on Vietnamese Secondary Forests" to Hà Nội University. Although he was approaching retirement, Minister of Higher and Secondary Education Tạ Quang Bửu created conditions for him to carry out this research and study for his Habil Doctorate in Dresden. He was awarded a four-year scholarship, but after spending a year to improve his German, he completed his habil doctorate in twenty months.

The president of his dissertation jury, Horst Kurtz, wrote to the Vietnamese Minister of Forestry, "In the name of the jury, I am happy to inform you that Mr. Nguyễn Văn Trương is qualified for the title of Habil Doctorate at the Technical University at Dresden. He is the first foreign researcher to be granted this title at our university."

When he was back in Hà Nội, Nguyễn Văn Trương happened to meet an old classmate from the Indochina College of Agriculture and Forestry during the French colonial period. His classmate, thinking that his early return meant he had failed to complete his degree, consoled him, "At our age of retirement, it's no wonder that we fail. It is not easy to pass between the rocks of modern science."

He smiled and said, 'Thank you for your sympathy."

Productive in Old Age

His colleagues often call him by the nickname "the forester of Kiều," since he knows long passages of the versified novel *The Tale of Kiều* by heart. In addition to it, he also knows many other poems and lines from classical and contemporary Vietnamese and French literature.

He particularly likes to cite a poem by Nguyễn Công Trứ, a renowned nineteenth-century scholar, economist, and administrator:

"Born on the earth and under the skies
You must engrave your name on the mountains and rivers."

"Mountains and rivers" here signifies the nation. For the sake of modesty, he has changed the second line to, "You must do your share for the welfare of your country."

This motto has no doubt inspired him throughout his life even during his elderly years. He is now eighty years old. Since his "retirement" at sixty, he has continued to work to the best of his ability.

First, he was appointed Advisor to the Minister of Forestry. In 1982, the ministry formed a plan for national reforestation. Faced with an alarming level of forest destruction, he directed a large-scale experiment using his theory of tri-dimensional forest structure. He used Nghĩa Đàn District, Nghệ An, as the pilot location for a model to be adopted from south to north. After ten years, species such as poplar, apricot, and teak had grown sufficiently for harvesting.

In 1990, as mentioned above, he founded the private Institute of Ecological Economics. At the beginning, many people felt doubtful about the institute's viability. But the results of thirteen years of activity have dispelled even the most pessimistic forecasts.

The IEE has created ten "ecological villages" in different natural regions. In addition to the Dao cooperative on Ba Vì Mountain, there are four villages in sand-dune environments (three in Quảng Bình and one in Quảng Trị). Hải Thụy (Quảng Bình Province) was famous during the American air war due to its women-only militia who shot down an American Phantom. One member of IEE, a veteran, came to Hải Thụy and taught people there how to dig irrigation wells under the sand. The water from these wells is used to grow mango trees, coconut palms, green vegetables, corn, and sweet potatoes. Trees were planted to set up a barrier against sand storms, and ponds were dug for raising fish.

The ecological village of Vinh Hòa (Quảng Trị Province) has been so successful that it currently attracts tourists. Not long ago, many residents of this village had to migrate elsewhere when sand storms hit their fields. He has invited several of these villagers to return home. One example is Mai Đình Đàn. When he came back to Vinh Hòa, Mai Đình Đàn built a thatched cottage, now replaced by a multi-story brick house. His field is full of watermelons, his pond is full of fish, and his gardens grow endless gourds and pumpkins. Fifty-three households in Vinh Hòa have transformed 142 hectares of sand into arable land; bred 150 oxen, 300 pigs and thousands of poultry; and planted 300 hectares of trees as a barrier against the sand.

In the flood zone, the institute chose Phú Điền (Hải Dương Province) as a pilot village. Its 100-hectare rice field yields only one rice crop per year. For the rest of the year it is under water. Under IEE's model of land use, one quarter of each household's land is used for digging a fish pond. The dug-up earth is then used to make a garden and an orchard in another quarter. The remaining half of the land is still reserved for rice planting using high-yield hybrid seeds to ensure sufficient rice. With this readjustment, farmers' income has increased five times.

Xuân Lâm Village (Thanh Hóa Province), with an area of 2000 hectares, was chosen as a pilot village for estuary regions. Its brackish-water mangroves reach a height of 1.5

meters. Ponds were to be dug in 10% of the total area and the dug-up earth would have served to reclaim a plot of land for planting fruit trees. The work was on the right track. Green vegetation began coming to life again, and birds and crabs began returning. But the villagers had to stop implementing the plan due to a lack of funds.

Other ecological villages are also being built. One is an ethnic Tày village in Hòa Bình Province. Another is Bình Dương Village (Vĩnh Phúc Province) for the preservation of medical plants. Tiên Điền Village (Hà Tĩnh Province) is the birthplace of poet Nguyễn Du, who wrote *The Tale of Kiều*. Another ecological village is located in the highlands of Đắc Lắc.

In March 2003, I had the pleasure of accompanying him and some of his colleagues to Yến Lạc, a mountainous district town in Bắc Cạn Province, over 150 kilometers north of Hà Nội. The last section of the road meandered between steep cliffs and ravines. In the distance over the trees on denuded slopes were floating clouds of smoke. He remarked with a sad voice, "Despite our efforts to make mountain cultures more stable and profitable, people go on practicing slash-and-burn farming. There's nothing to do about it, a thousand-year tradition can't be changed overnight. An old proverb says, *'Đao canh hỏa chủng'* – clear the fields with a machete, then burn them for planting. With the accumulation of topsoil, the burned fields are fertile for a few years. But erosion, especially by rain water, soon exposes the lower layers of gravel and late rice, which are infertile. Then people leave their fields on the slopes to go and burn the forest in another place. It is very difficult to grow plants on hills, that is why they keep on using slash-and-burn methods even after adopting a stationary lifestyle."

IEE works to end slash-and-burn farming with the Tày ethic group in Na Rì by carrying out a project of intensive sloped-land agriculture. He and two other professors came to Kim Lư Village to conduct a training course for forty households. IEE teachers taught agronomic techniques. One expert, Trương Quốc Long, stayed for a longer duration in the village to demonstrate methods of combating erosion and retaining topsoil, controlling flash floods, and cultivating fast-growing plants to improve the soil. He also showed people in Kim Lư new ways to dig drains and wells, to set up obstacles to slow down flood waters, and to prevent the land from eroding. To achieve this, IEE paired Trương Quốc Long with two commune officials, the vice president of the People's Committee and the vice president of the Women's Union. At the same time, two young university graduates in biology were sent to Ba Bể and Bắc Cạn to organize "ecological gardens" and to assist with a project for harvesting forest products (bamboo, rattan, and medicinal plants).

Finally, the octogenarian scientist Nguyễn Văn Trương did not limit his activities to his specialized fields of forestry and ecological economics. He was editor-in-chief of the 4000-page, four-volume *Vietnamese Encyclopedia*. He was also a jury for state prizes awarded to distinguished scientists. An indefatigable and strongly motivated worker, he is truly enjoying a productive retirement.

At the age of eighty-four years old, he was awarded the title "Hero of Labor during the Renovation Period" by the Vietnamese government.

NGUYỄN TÀI CẨN: ENLIGHTENING THE ORIGIN OF VIETNAMESE

Being the first person in Việt Nam to successfully defend his Ph.D. thesis in linguistics in the former Soviet Union, Nguyễn Tài Cẩn also became the first linguist to receive the Hồ Chí Minh Award in 2000. His research aims to resolve problems about grammar and phonetics in the Vietnamese language, the origin of the pronunciation of Sino-Vietnamese, the Nôm (Chinese-transcribed Vietnamese) script, and *Truyện Kiều* (The Tale of Kiều).

In 2006, on his eightieth birthday, the Việt Nam National University in Hà Nội presented to him a big white-glazed porcelain plate with handwritten Chinese characters in indigo-blue glaze. They featured shortened words from *Luận ngữ* (The Analects of Confucius), "Never get fed up with studying. Never get tired with teaching." That description is absolutely in line with his characteristics.

Which Way to Go in the Middle of Youth?

Nguyễn Tài Cẩn studied at the same school (*Quốc học Huế* or Huế National School) as Hoàng Tụy, Nguyễn Văn Hường, Nguyễn Hoàng Phương, and Nguyễn Pháp. (These men later became famous professors in Hà Nội.) He was good at all subjects, but best in French. He nourished his ambition to become a writer in French. On his visit to his former teacher, Mr. Hoài Thanh, Cẩn told him about this ambition. Unexpectedly, Mr. Hoài Thanh said, "We are Vietnamese and are learning French in a French protectorate. How can we write in French as well as French writers in their own country?"

After that, he decided to study Vietnamese and built a passion for composing poems in Vietnamese. One day, when visiting his former teacher, Mr. Đào Duy Anh, Cẩn asked for his opinions about the poem collection Cẩn had composed. He had hoped to receive compliments from his teacher. To his surprise, the teacher said deliberately, "If you have an endowed talent, you can write good poems. If not, it's just a waste of effort doing so. In contrast, if you are persistent and have good methods, you'll discover many surprising and useful things and become a good scholar."

The teacher's advice helped him orient his career. He was determined to become a great scholar, not just a normal poet.

Contributing to Building Vietnamese Grammar

Up until the 1960s, many scholars still applied the grammatical structures of foreign languages into Vietnamese. According to them, the sentence "Chiếc lá này xanh" that is naturally uttered by Vietnamese students is incorrect grammatically and, therefore, should be corrected into "Chiếc lá này *là* xanh" as seen in English grammar (This leaf *is* green) or French grammar (Cette feuille *est* verte). However, each language has its own distinctive features. These scholars should have analyzed the Vietnamese language and worked out rules for Vietnamese grammar instead of simply applying foreign grammar to this language.

It is not hard to imitate foreign grammar. On the contrary, it is much more challenging and difficult to carry out thorough surveys on the different uses of the Vietnamese language and discover its internal rules. Nguyễn Tài Cẩn took this approach right from the start.

In regards to the book *Ngữ pháp tiếng Việt: tiếng, từ ghép, đoản ngữ* (Vietnamese Grammar: Words, Compound Words, and Phrases) by Nguyễn Tài Cẩn, Cao Xuân Hạo, a linguist, had a review published in the *Lao Động Newspaper* on 8 September 2000. He wrote, "Forty years ago, Nguyễn Tài Cẩn, a scholar of Vietnamese studies, introduced a theory on syllabi and morphemes from the Soviet Union that helped modern linguists solve problems related to such an isolating language as Vietnamese and finally find ways to overcome 'Europeocentrism.' Unfortunately, no one understood him then."

Cao Xuân Hạo had made a lot of significant achievements in Vietnamese phonetics, grammar, and semantics. Therefore, his evaluation of Nguyễn Tài Cẩn was very reliable.

Swimming Upstream for Centuries

Sino-Vietnamese vocabulary makes up a large percentage of Vietnamese vocabulary. It accounts for about 60-70 percent of argumentative writings. Its uses can be considered unlimited. Vietnamese users can borrow new Sino-Vietnamese words when necessary without breaking the internal structure of the Vietnamese language. Not only old scholars but also modern scientists who are fluent in Sino-Vietnamese, such as Professor Hoàng Xuân Hãn, have enriched Vietnamese with modern Sino-Vietnamese words such as *định lý* (theorem), *định luật* (law), *đẳng thức* (equality), *phương trình* (equation), *nguyên tử* (atom), *điện tử* (electron), and *lượng tử* (quantum). However, Vietnamese people say *phương trình vi phân* instead of *vi phân phương trình* in accordance with Chinese grammar. Similarly, Vietnamese ancestors said *hoa mai* instead of *mai hoa* or *ngoại thành* instead of *thành ngoại.*

Many Sino-Vietnamese words are used in political, military, and diplomatic fields. Examples include *duy vật* (materialism), *duy tâm* (idealism), *cương lĩnh* (political program), *chính sách* (policy), *chiến lược* (strategy), *sách lược* (tactic), *du kích* (guerilla), *chính quy* (regular), *tiến công* (attack), *phòng ngự* (defend), *sư đoàn* (division), *quân đoàn* (corps), *tả khuynh* (left wing), *hữu khuynh* (right wing), *giáo điều* (dogma), and *kinh viện* (scholasticism).

How can such a great number of Sino-Vietnamese words be integrated into Vietnamese without changing the Vietnamese language's internal structure? It is because Vietnamese people pronounce Chinese words in the manner of Sino-Vietnamese, not in the manner of *Quanhua* (Ancient Mandarin Chinese) or *Putonghua* (Standard Mandarin Chinese) based on the Beijing dialect of Mandarin Chinese.

A Big Question for Orientalists

Linguists of Vietnamese and Chinese, as well as international Orientalists, pay attention to the pronunciation of Sino-Vietnamese. The study of the pronunciation of Sino-Vietnamese helps scholars understand other aspects of medieval Chinese such as the pronunciations of Sino-Japanese and Sino-Korean.

According to historical documents from the Han dynasty, some Chinese governors such as Ji Guang, Ren Yan, and Shi Nie made an attempt to "sinicize" Giao Chỉ and Cửu Chân (districts in Northern Việt Nam under Chinese domination). Shi Nie was later honored as "the father of studies in the South" by Vietnamese historians.

Recently, the temple and tomb of Shi Nie in Luy Lâu (Bắc Ninh Province) have been recognized as historical-cultural relics by the Vietnamese government. However, it wasn't until the Sui and Tang dynasties that scholars of Giao Châu (Northern Việt Nam under Chinese domination) could be compared with those of Trung Nguyên (the land between the Huang He/Yellow River and Yangtze Rivers in China) in terms of knowledge of sinology. Khương Công Phụ, a native of Cổ Hiểm Village, Cửu Chân District, Ái Châu Province (present-day Yên Định District, Thanh Hóa Province), went to the capital of Chang'an (present-day Xi'an), China, to take part in the highest imperial and metropolitan examinations. He passed all the examination and became *Trạng nguyên* (Number One Scholar). He was invited to stay in Chang'an by the emperor of the Tang dynasty and was later appointed as *Tể tướng* (Prime Minister) in China. Under the Tang dynasty, the number of people that went from An Nam (Northern Việt Nam under Chinese domination) to China to take the metropolitan and highest imperial examinations was limited to eight for each examination. The temple dedicated to Khương Công Phụ is still in Thanh Hoá Province, Central Việt Nam.

Most of Orientalists agree that the current pronunciation of Sino-Vietnamese originates from the pronunciation of Chinese under the Tang dynasty in the capital city of

Chang'an. It is based on the phonetic system of Chinese which was last taught in Giao Châu before Việt Nam gained its independence in the tenth century. Since the tenth century, the pronunciation of Sino-Vietnamese has developed in accordance with Vietnamese and in separation from Chinese. The Sino-Vietnamese vocabulary has been integrated more and more deeply into the Vietnamese vocabulary. Present, Sino-Vietnamese is no longer an "implanted" but integral part of the Vietnamese language.

Looking Back Ten Centuries Through Hundreds of Books

To reach conclusions, Nguyễn Tài Cẩn had to deal with a lot of issues. For example, if the present pronunciation of Sino-Vietnamese stemmed from the pronunciation of Chinese characters in the Tang dynasty in Chang'an (Tchang-an), then how did Chinese people in the capital city pronounce those characters? What consonants, rhymes, and tones were included in Chinese at that time? What international phonetic symbols can be used to transcribe Chinese? These questions cannot be answered without solid foundations.

Moreover, how did Vietnamese people pronounce Chinese characters in the eighth and ninth centuries? What consonants, rhymes, and tones were included in Chinese during these centuries? How did the pronunciation of Sino-Vietnamese change from the tenth to twentieth centuries? Only by using modern methods of historical phonetics can one answer these questions.

Books such as *Le dialecte de Tchang-an sous les Tang* (The Chạng'an Dialect Under the Tang Dynasty) by H.

Maspéro, *Études sur la phonologie chinoise* (Studies on Chinese Phonology) by B. Karlgren, and *Thiết vận* by Lu Fa Yan in Chang'an in the early seventh century were important reference materials for Nguyễn Tài Cẩn.

He also read many other books by Orientalists in Chinese, Nôm, Vietnamese, English, French, German, Russian, and Japanese about related issues. With the knowledge drawn from these books and with his own arguments and interpretations, he shed light on the origin of the pronunciation of the Vietnamese language that helped Vietnamese people receive the influence of the Chinese culture without being "sinicized."

Learning About the Vietnamese Phonic System from Four Thousand Years Ago

The book entitled *Lịch sử ngữ âm tiếng Việt (sơ thảo)* (History of Vietnamese Phonetics (draft) by Nguyễn Tài Cẩn published in 1995 was included in the "project" that earned him the Hồ Chí Minh Award. Through the 350 pages of the book, he presented clearly the development of Vietnamese phonetics from the primitive age until the present day.

Nguyễn Tài Cẩn and his students critically applied the Swadesh method, a grotto-chronological method. It is based on the conception that the replacement of an old basic vocabulary with a new one over times is the common rule of all languages. If two languages or dialects of the same origin are separated from each other for about 1000 years, about 74 percent of the original 100-word vocabulary list will still be shared by them. If they are separated for 2000 years, that number will be about 54 percent. And if they are separated for 4000 years, that number will be around 30 percent.

Although it is called modestly a "draft" by its author, it is actually a ground-breaking work. To complete it, he had to refer to hundreds of studies on Vietnamese, Mường (a language spoken by the Mường people in Việt Nam), and other related languages such as Nguồn, Poọng, Mày, Rục, Sách, Mã Liềng, A-rem, Thà Vựng, Pakatan, Phon Soung, and Khạ Phong. Nowadays, it is widely agreed that Vietnamese belongs to the Southern Asian language family, the Môn-Khmer branch, and the Việt-Chứt line. It is in contact with the Thai-Kadai and Chinese languages but not closely related to them.

In order to have a good understanding of the Vietnamese phonetics from the seventeenth century, he referred to *Dictionarium Annamiticum-Lusitanum-Latinum* (Vietnamese-Portuguese-Latin Dictionary) by Alexandre de Rhodes published in 1651 in Rome, Italy.

For Vietnamese phonetics in the fifteenth and sixteenth centuries, he read *An Nam dịch ngữ*. In this book, more than seven hundred Vietnamese words were translated and transcribed into Chinese characters by Chinese people for diplomatic purposes.

Mường – The "Twin" of the Vietnamese Language

More than one thousand years ago, the Vietnamese and Mường peoples used the same language. Nguyễn Tài Cẩn is from the Thanh Chương District, Nghệ An Province, and speaks Vietnamese with a Nghệ An accent. He is also fluent in the dialects of Zone 4. This is very useful for his study of the ancient Vietnamese language because the Nghệ An dialect still maintains many of its sounds.

Nguyễn Văn Tài, his student and colleague, made a thorough investigation into the twenty-nine dialects of the Mường language and reported the results in his Ph.D. thesis in 1983. These dialects were categorized into nine groups scattered in the north and center of Việt Nam. These groups included Mường Phong, Mường Thái, Mường Bỉ, Giai Xuân, Tân Hợp, Sông Con, and others.

The book written by Nguyễn Tài Cẩn has helped lovers of the Vietnamese language, including writers and journalists, better understand Vietnamese and use it in a more correct manner. After reading the book, hardly anyone can say that they understand Vietnamese thoroughly.

Some of his students such as Nguyễn Văn Bá and Ngô Thị Giáng Hương used the Swadesh method to compare the Mường language in the mountainous area of Nghệ An and the Vietnamese language in the territorial waters of Nghệ An (such as Nghi Lộc). They found out that the rate of common vocabulary came up to 85.5 percent. Meanwhile, the Vietnamese dialect in Hà Nội and Nghệ An shared a lower rate, about 82.2 percent. It is said that the Vietnamese dialect in Nghệ An is closer to the Mường language than the Vietnamese dialect in Hà Nội.

Looking for the Source of Consonants and Vowels

He carefully looked for the source of the initial consonants (such as nasals, occlusives, and fricatives), front vowels (e, ê, i, and iê), central vowels (a, ă, ư, ơ, â, and ươ), final sounds (m/p, n/t, ng/c, u/o [w], i/y [j]), and the schwa "w". He then looked for the source of tones. He analyzed and combined the findings of his colleagues both domestically and overseas through field trips or judgments made by many world-famous linguists. For example, M. Ferlus not only took a thorough study in the pre-Vietnamese-Mường language but also did a survey on the Vietnamese dialect in Vinh, Nghệ An Province. M. E. Barker studied the Mường phonemes as well as the final consonants and vowels of the pre-Vietnamese-Mường language.

To write the 350-page course-book, he had to refer to an overwhelming abundance of materials, assess them, and propose his own interpretations.

Still Lacking the Version of *Truyện Kiều* Which is Closest to the Original

On returning to Tiên Điền Village, the "Land of Culture" in Nghi Xuân District, Hà Tĩnh Province one day, I visited the Monumental House of Nguyễn Du. The residents here consider characters in *Truyện Kiều* by Nguyễn Du as their close relatives. *Truyện Kiều* has lived in the heart of Vietnamese people throughout the country for the last two centuries and has been expressed in many original forms, such as *chèo* Kiều (traditional operetta of Kiều), *vịnh* Kiều (composing poems of Kiều), *tập* Kiều or *lẩy* Kiều (combining verses of the same rhyme selected from *Truyện Kiều* to make up a description of a thing or a happening), *bói* Kiều (telling fortunes based on *Truyện Kiều*), and others.

It is a pity that there is still a lack of the version of *Truyện Kiều* that is closest to the original. There are many reasons for this. First, the Nôm scrip was underestimated in Việt Nam under the monarchy. Confucian scholars at that time were taught and took exams in Chinese, but not in the Nôm script. The Nôm script stemmed from Chinese characters, and it was transcribed differently by different people. As a result, modern scholars have to read, guess and "decode" writings in Nôm at the same time.

Second, there was not any official Nôm dictionary as the standard for national documents in Việt Nam under feudalism. In contrast, Chinese dictionaries were published in China more than two thousand years ago. They increased in number and became richer and more accurate in content later.

Third, ancient Vietnamese carvers were not used to creating wood blocks in the Nôm script. Therefore, the text of *Truyện Kiều* was sent to China to be carved by workers in Guangdong Province. Because these Chinese workers could not read the Nôm script, they sometimes spontaneously turned parts of the text they thought were incorrect into Chinese. Therefore, handwritten and printed versions of *Truyện Kiều* unavoidably included a lot of mistakes when compared with the original.

Also, the first versions of *Truyện Kiều* in Vietnamese were based on different versions of the Nôm script and, therefore, strayed further and further away from the original.

***Truyện Kiều* has Been Honored, But...**

After the August Revolution in 1945, Vietnamese became the official language of Việt Nam, and literature in Vietnamese played an important role. *Truyện Kiều* was taught at all educational levels and studied at institutes and universities. Nguyễn Du was recognized as a World's Cultural Celebrity. The revision of this literary work was paid more attention to than ever before. The work *Từ điển Truyện Kiều* (Dictionary of the Tale of Kiều) by Đào Duy Anh was introduced to readers in 1974.

In spite of noticeable advances in the revision and study of *Truyện Kiều*, scholars still have to do many things to restore its original version. From 1982 to 1990, Professor Nguyễn Tài Cẩn was twice sent to Paris University 7 where he gave lectures for about two years. He met and exchanged ideas with Professor Hoàng Xuân Hãn, who had spent almost half a century studying *Truyện Kiều*. This professor advised him to search for and study as many as possible ancient versions of *Truyện Kiều* in the nineteenth century in order to obtain enough scientific grounds for the restoration of its original.

A Breakthrough in the Study of *Truyện Kiều*

While working in Paris, Professor Nguyễn Tài Cẩn made a lot of effort in searching for different versions of *Truyện Kiều*. With the support from his friends, he finally collected nine copies from the nineteenth century. However, he did not start writing immediately because he was still waiting for the book *Kiều tầm nguyên* (Seeking the Original of the Tale of Kiều) by Professor Hoàng Xuân Hãn to be introduced to readers.

However, many years have passed but that book was not published. Therefore, he decided to release his 645-page book entitled *Tư liệu Truyện Kiều: Từ bản Duy Minh Thị đến bản Kiều Oánh Mậu* (Materials on The Tale of Kiều: From the Version by Duy Minh Thị to the Version by Kiều Oánh Mậu) published by the Literature Publishing House in collaboration with the Center for National Culture Studies.

In the preface and conclusion of the book, he used very modest words to talk about his work, "While the book *Kiều tầm nguyên* by Professor Hoàng Xuân Hãn has not come out yet, readers can temporarily use this book. …When that book is published, we will have chance to make comparison. Although the two books have the same target, there are certainly differences. ...It is possibly because we just want to add some more linguistic knowledge to what is presented by him."

The book by Nguyễn Tài Cẩn was considered a breakthrough in the study and revision of *Truyện Kiều* in the early twenty-first century. It is noticeable that he used modern scientific methods in revising and studying this work.

In his recently-published book, he presented a thorough survey of nine versions of *Truyện Kiều* in the nineteenth century. They include seven versions by Duy Minh Thị (1872, 1879), Liễu Văn Đường (1871), Quảng Văn Đường (1879), Thịnh Mỹ Đường (1879), Lâm Noạ Phu (1870), and Kiều Oánh Mậu (edited in the late nineteenth century and published in 1902); one version, which is half in the Nôm script and half in Vietnamese, by Abel des Michels (1884); and the other, which is in Vietnamese, by Trương Vĩnh Ký (1875).

Investigating the nine different versions of *Truyện Kiều* meant surveying about 29,000 verses or around 205,000 words. Most of them were handwritten or printed manually more than five hundred years ago, so they were dim and difficult to read. Moreover, some included ancient characters, so they are hard to "decode."

After the book had been introduced, he still continued to study the versions of *Truyện Kiều* in the ancient Nôm script. He discovered some details that had been considered taboos in the Lê-Trịnh period (1533-1788). It is inferred from this discovery that Nguyễn Du might have written *Truyện Kiều* in the Tây Sơn period (1778-1802) during his ten years of living away from his hometown, not after his return to Việt Nam from China as thought by many scholars of *Truyện Kiều*.

Although he was trained in Russia and got married to a Russian woman, Professor Nona Stankyevitch, he has spent most of his life studying the Chinese-Nôm language and had many works published in this field. Some of them included *Nguồn gốc và quá trình hình thành cách đọc Hán-Việt* (The Origin and Formation Process of the Pronunciation of the Chinese-Vietnamese Language), *Một số vấn đề chữ Nôm* (Several Issues On the Nôm Scrip), *Ảnh hưởng Hán văn Lý-Trần qua thơ và ngôn ngữ thơ Nguyễn Trung Ngạn* (Impact of the Chinese Culture in the Lý-Trần Period on the Poetry and Poetic Language of Nguyễn Trung Ngạn), *Giáo trình lịch sử ngữ âm tiếng Việt* (A course-book On the History of Vietnamese Phonetics), and *Tư liệu Truyện Kiều: Bản Duy Minh Thị* (Material on the Tale of Kiều: The Version by Duy Minh Thị). After retiring, he and his wife moved to Moscow to be taken care of by their children on days of illness. Still, they are absorbed in studying the Vietnamese language as they did during the years of teaching at Hà Nội University and evacuating in the forest and mountainous area of Đại Từ, Thái Nguyên Province.

HOÀNG TỤY AND THE THEORY OF GLOBAL OPTIMIZATION

Ann Hibner Koblitz, an American history professor, founded the Kovalevskaya Prize for outstanding female scientists in Việt Nam, Nicaragua, and El Salvador. Her husband, Neal Kobliz, is a professor of mathematics at the University of Washington. When he returned home from Việt Nam in 1990, he wrote a 30,000-word article for the prestigious review *The Mathematical Intelligencer*, entitled "Memories of Mathematics in an Isolated Country." The major part of this article is devoted to the life and the work of Hoàng Tụy, the leading figure of the Hà Nội School of Mathematics, whose theory of global optimization has become a classic and is recognized around the world.

Hoàng Tụy was born on December 27, 1927. In 1997, on the occasion of his seventieth birthday, a three-day workshop was organized in Linkoping (Sweden) addressing the progression from local to global optimization. The presentations at this conference were collected in a publication in honor of Hoàng Tụy and published by Kluwer Academic Publishers. According to this book, Hoàng Tụy is considered "the father of global optimization." Numerous technical terms such as "Tụy's cut," the "Tụy-type algorithm," and "Tụy's inconsistency condition" have been named after him. In 1996, Hoàng Tụy was given the Hồ Chí Minh Award for his work on optimization.

Hoàng Tụy came from Xuân Đài Village in the central province of Quảng Nam. He is descended from a family of patriotic Confucian scholars, of which the greatest glory came to his great-uncle, Hoàng Diệu, the military governor of Hà Nội who committed suicide in 1882 instead of surrendering to the French invaders.

When Hoàng Tụy was young, he demonstrated a particular gift in mathematics. However, a grave ill forced him to give up his studies at the age of fifteen. Recovering from this disease thanks to a tratraditional doctor, he returned to school in Huế, though still in poor health. After private studies, he succeeded in earning his baccalaureate in mathematics in 1946, one year after the revolution. With a small sum of money he had earned as a tutor, he went to Hà Nội in the hope of studying with the eminent mathematician Lê Văn Thiêm. Lê Văn Thiêm was the first Vietnamese to earn a doctorate in mathematics, and he was expected to return to Việt Nam from abroad.

However, the Franco-Vietnamese war that broke out in December 1946 prevented his return. The university closed its doors. After buying all the interesting mathematics books he could find in the capital, Hoàng Tụy went back to Quảng Nam. Ever since his father had died when he was four, he had lived with his mother, brothers, and sisters in a mountainous region. Here, he taught in a private junior high school from 1947 to 1951, while continuing to study math on his own. During this period, his school was bombed by French Spitfires, and the community grieved over the death of one teacher and seventeen students.

Taking a correspondence course, he was able to receive a certificate for general mathematics. Learning of Lê Văn Thiêm's return to Việt Nam and the formation of the College of Fundamental Sciences in the northern mountainous region of Việt Bắc, Hoàng Tụy decided to go there. He had to walk hundreds of kilometers along a part of the path that was later known as Hồ Chí Minh Trail. During this journey, he had to face three continual dangers: French ambushes, tigers, and malaria. Along the way, he stopped for some times in the liberated Interzone 4 where he gave mathematics lessons in order to earn a little money to continue his journey. He finally arrived in Việt Bắc in September 1951.

At the College of Fundamental Sciences, however, all of the courses were under his level. He had to give private lessons while deepening his mathematical knowledge on his own. Fortunately, he was finally able to meet his idol, Professor Lê Văn Thiêm.

With his reputation as a good secondary school mathematics teacher, he was transferred to the temporary Vietnamese Educational Center outside Nanning, China, where there was no danger of French bombardment. There he could have access to high-level Russian mathematical books. He learned Russian in order to decipher these books.

After the liberation of Hà Nội in 1954, schools of higher learning reopened. In September 1955, Lê Văn Thiêm, by then Director of the College of Sciences, invited him to come and teach at the college. In July 1957, he, along with eight other professors, was sent to Moscow for a study course. He went to work at the Department of Mathematics and Mechanics at Lomonosov University and began his career as a researcher.

He chose real analysis as the subject matter of his research. He worked under the direction of Professors D.E. Menshov and G.E. Shylov. He was nervous that the professors would not have much confidence in an autodidact like him. Fortunately, he was able to answer a question Shylov asked him about the Lebesgue Measure with an original demonstration. This won him Shylov's respect. In 1959, a year and a half after his arrival in Moscow, he successfully defended his doctoral thesis under the direction of academician Menshov.

On his return to Hà Nội, he was appointed Dean of the Department of Mathematics of Hà Nội University. He went on studying real analysis and published five works on this subject in the Russian mathematical magazine *Mat. Sbornik*. Nevertheless, he realized that this purely theoretical field could not offer useful applications in Việt Nam. Having worked for a while in 1962 with L.V. Kantorovich, a famous Soviet mathematician, economist, and Nobel Prize winner in 1975, and studied Chinese documents, he determinedly turned to operations research. In 1963-1964, he came into contact with several Chinese mathematicians, however the cultural revolution in 1966 put an end to these relations.

Before 1964, although optimization had been a subject of research, attention was only centered on local optimization whereas the problems of global optimization were considered intrinsically and inherently difficult. At a 1964 seminar held by Kantorovich in Novosibirsk, Hoàng Tụy presented a solution to a basic problem in global optimization. This solution, published in a report of the Academy of Sciences of the USSR, has become a classic and is known as "Tụy's Cut."

In 1964, the Americans began to bomb North Việt Nam. Hà Nội University was evacuated to the mountainous region of Thái Nguyên, 170 kilometers north of Hà Nội. The Department of Mathematics and Physics directed by Hoàng Tụy comprised of 250 students. The classrooms were in bamboo cottages sheltered under high trees. In the first minute of an air raid alert, everyone jumped into trenches dug under the bamboo tables.

In 1965, the Vietnam Mathematical Society was founded, with Lê Văn Thiêm as president and Hoàng Tụy as general secretary. It held regular twice-monthly meetings in Hà Nội for professors from three schools (the Hanoi University, the Teachers' training College, and the Polytechnic University) that had each been evacuated to different locations. In 1967, during the height of the American air escalation, the Fields Medal-winning French mathematician Alexandre Grothendieck gave lectures to the society on category theory, homological algebra, and algebraic geometry. After a bomb exploded less than a hundred meters from the lecture hall, the following sessions had to be organized in an evacuation area. The French professor thought highly of the professional level of young Vietnamese mathematicians and proposed to send some of them to France to continue their researches.

In August 1969, a month before his death, President Hồ Chí Minh invited Hoàng Tụy to come and see him. The President wanted Tụy to find a solution through operations research to relieve crowding in the numerous beer halls in Hà Nội. The long lines that formed in these establishments were dangerous in time of bombing raids. However, this appointment had to be cancelled because of the president's state of health.

It was Prime Minister Phạm Văn Đồng who, several weeks later, received the mathematician. Because of heavy traffic, Hoàng Tụy arrived a little late. After sitting near the Prime Minister, he realized that President Hồ Chí Minh was also present among the numerous officials in the room. President Hồ Chí Minh asked him, "Can't you find a different term than the Sino-Vietnamese *vận trù* to translate "operations research?" Something simpler that everyone can understand?"

Hoàng Tụy answered that this term had been adopted by Vietnamese scientists since 1961 and enabled them to read Chinese materials on the subject more easily. Hồ Chí Minh explained that the original meaning of *vận trù* referred to war plans and strategies in classical Chinese novels.

A few months after the President's death, Hoàng Tụy was summoned by Prime Minister Phạm Văn Đồng again, who asked him to report on the results the operations research group had achieved on the decrowding of beer halls. There results were positive. Nevertheless, the bombing by US B-52s had upset everything.

In 1970, the Institute of Mathematics was created under the State Committee of Science and Technology. Lê Văn Thiêm became director of the institute, and Hoàng Tụy began working there. American bombing finally ended in 1973, and the war came to an end in 1975. In that year, during Prime Minister Phạm Văn Đồng's visit to the USSR, Kantorovich presented him with a book on economic optimization. Hoàng Tụy was charged with reading it and making a report to Phạm Văn Đồng.

In 1976, Hoàng Tụy was invited to give a presentation at the International Congress on Mathematical Programming which gathered some 800 mathematicians in Budapest. On this occasion, French mathematicians invited him to Paris.

With eighty scientists, the Institute of Mathematics had been doing good work, especially in the field of real analysis. Lê Văn Thiêm and Hoàng Tụy's efforts were supported by Prime Minister Phạm Văn Đồng, Tạ Quang Bửu, and foreign scientists such as the Fields Medal-winning French mathematician Laurent Schwartz. Professor Hoàng Tụy directed a competent team engaged in research on optimization.

The weak point of the institute was the Department of Applied Mathematics, since it lacked modern equipment, sophisticated infrastructure, and an advanced industrial base. During the anti-American war, Hoàng Tụy received the support of world-renowned foreign mathematicians

Seventeen mathematicians of the institute had worked at the Humboldt Institute in the German Democratic Republic. In 1973, Hoàng Tụy was in Moscow during the International Mathematic Olympiad for secondary school students. On his suggestion, his East German colleagues helped Vietnamese students to take part in the subsequent competition in 1974. Since then, Vietnamese participation in the Olympics has become routine, and Vietnamese students have generally achieved excellent results.

"Tụy's Cut" had received attention from American mathematicians as early as 1964. At a 1976 conference in Budapest, a mathematician from the NEC Company told Hoàng Tụy that his research group had successfully applied this method to a difficult problem.

Hoàng Tụy's works constitute a landmark in the birth of global optimization theory. After the signing of the 1973 Paris Agreement stipulating the suspension of American bombing of North Việt Nam, he was appointed Vice President of the Council of Sciences and then Director of the Institute of Mathematics. He continued to gather and train young mathematicians and publish his personal works.

In 1981, together with Lê Văn Thiêm and Nguyễn Văn Đạo, Hoàng Tụy visited America, where his work had become well known. Since 1976, his name had appeared on the editorial committee of the journal *Mathematical Programming*, but it was often mistaken for a Chinese name. He also participated in the editorial committee of international journal *Optimization*. At the time that he visited America, Professor Stephen E. Jacobsen had just published an article on Tụy-type algorithms in the journal *Applied Mathematics and Optimization*.

In the 1980s, the theory of global optimization developed on a worldwide scale, bringing the Hà Nội School of Mathematics a global reputation. On the basis of his own research and that of Vietnamese and foreign colleagues, Hoàng Tụy collaborated with the German Reiner Horst to publish the 694-page English book *Global Optimization: Deterministic Approaches* (Springer-Verlag, 1990). In 1988, the Thirteenth International Conference on Mathematical Programming decided to create a new journal, *Global Optimization*, and invited Hoàng Tụy to become its editor-in-chief. He refused because the means of communication in Việt Nam were then still behind the times.

In the early 1990s, he occupied himself with D.C. programming, measuring the difference between two convex functions. In 1996, he published the book *Optimization of Low Rank Non-Convex Structures* (Kluwer Academic Publishers) along with Japanese Professor

Hiroshi Konno and the young Vietnamese mathematician Phan Thiên Thạch. As the author of 140 works in all, he was presented with an honorary doctorate from the University of Linkoping in Sweden. His collected lectures have provided material for a classic doctoral manual (Kluwer Academic Publishers, 1997).

NGUYỄN VĂN HƯỜNG: CROSSING RIVERS UNDER AMERICAN BOMBS

Nguyễn Văn Hường was born in 1927 in Nam Đàn District, Nghệ An, the home province of Hồ Chí Minh. He descended from a family of patriotic scholars. He left school to join the army at the beginning of the resistance war against the French, who came back to reconquer Việt Nam one year after the 1945 Revolution had put an end to the eighty-year colonial occupation.

He served in the headquarters of the Central Highlands (Tây Nguyên) in southern Việt Nam. Life was hard. Because of a shortage of rice, the soldiers had to go to the forest and dig up wild tubers to eat. When salt became scarce, they had to resort to burning straw, whose ashes have a salty taste. They thrashed tree bark until it became soft enough to serve as blankets they needed against the biting cold of the highlands. But the comfortable and happy hours they spent by the hearth of ethnic minorities' stilt-houses left exciting and unforgettable memories.

The resistance needed brain-power and, therefore, young intellectuals needed to be trained. He was called to the mountain of Việt Bắc in northern Việt Nam, a stronghold of the revolution and resistance. To reach this zone, he had to walk for several months, covering a long distance along the path later known as Hồ Chí Minh Trail. He was assigned to the Arms and Ammunition Unit directed by the competent engineer, Trần Đại Nghĩa, who had graduated from universities in France. Nguyễn Văn Hường contributed to manufacturing the first batch of steel in Việt Bắc using handicraft-like methods.

He later continued his technical studies in China and the Soviet Union, where he succeeded in earning his doctorate six months ahead of schedule in order to return and serve his country. The war against the Americans had been underway for several years at that time. His thesis supervisor, academician Rzhamitsyn, embraced him on the day when the examination results were announced and told him, "When you are back in your country, save some time to deepen your knowledge of the questions that you have left unanswered. This will be an original contribution to the structural mechanics. Don't let yourself get lost in everyday matters!" He would never forget this recommendation.

Back in Hà Nội, he requested to be posted to the southern battlefront. However, he was assigned to his old school, the College of Civil Engineering, to join a group of researchers on structural mechanics. The researchers welcomed him like a teacher.

What could he do to serve the resistance? He went and knocked on the door of the Ministry of Transport and Communication, because the most important problem to solve was the transfer of troops, ammunition, and food to the southern front. The American Air Force attacked the supply lines of North Việt Nam without letup, chiefly aiming at ports and bridges. American reconnaissance planes searched for any sign of repair of destroyed bridges. All the bridges were collapsing under the bombing.

The solution was to develop a kind of bridge that could not be easily detected and could be taken apart and moved quickly so that Vietnamese troops could cross rivers in the flood season. Ordinary pontoon bridges were easily swept away by the swirling waters, and fords were unusable when the water rose quickly. People had heard of cable or rope bridges, but all the information available about them was contained in a few popular-science articles.

He was assigned to construct an experimental prototype. First, he had to calculate the length of the cables and the piers on either bank. The difficulty lay in the centrifugal force created by the cables that replaced the roadway.

Finally, the first experimental bridge was brought to light. The bridge was passable by cars, but only with winches turned by hand. This made the crossing dangerously slow. Enemy planes could intervene at any moment. He proposed to use an internal combustion engine to turn the winches. But the design of a device for the transmission of force did not come within his experience. He needed to solve the problem quickly in order to meet the demands of the regions that were being bombed the most often.

Another urgent problem was raised: how could the engineers ensure the safety of cars running on the cables? When heavily laden, cars were subjected to lurch. One car was propelled into the water along with its driver, who was fortunately able to swim. Heated discussions took place to discover the causes of this failure. Were the cables too long? Was the pier placed incorrectly? Was the knot joining the cables wrong? One experienced specialist declared, "Nguyễn Văn Hường has violated a fundamental technical principle, a classic one, in cable transportation: the load must have its center of gravity over the cables, not under them."

After having checked his calculations and considered different hypotheses, Nguyễn Văn Hường concluded that the failure had been due to the elasticity of the cables. The calculations on the stability of a heavy load travelling on an elastic rope did not exist. He had to start research from scratch.

After spending several sleepless nights, he built another cable bridge. There were still two piers on the banks and two cables connecting them, but the winches were gone. In their place were wheel-rims screwed onto the cables. A military driver volunteered to carry out the experiment. He jumped into the driver's cab next to him. The seven-ton truck advanced on the rope. The authorities, soldiers, mechanics, and military engineers stood on the banks, holding their breath. The experiment was like a high-wire act the circus, and equally as dangerous! Sighs of relief and applause rose when the truck arrived at the other side of the bridge.

After having invented a method of manufacturing cable bridges, he was ordered by the Ministry of Transport and Communication to construct a series of bridges for Southern Việt Nam, a zone whose roads were attacked day and night by the U.S. Air Force. No traditional bridge could hold up. He and his technical colleagues were moved by stories of the daily heroism of military drivers, ferrymen, and engineers who had to work even in full daylight. Young women from the Vanguard Youth were also involved with the construction and rebuilding of roads.

One night, he and his team arrived at the banks of the Cà Tang River near Mụ Gia Pass, flowing between two steep cliffs. This place was a continual target of American bombers. There were no traces of life, the inhabitants having evacuated their stilt-houses to seek shelter in the neighboring forests. It had been raining in torrents for several days. The river rose with lightning speed. He was shocked when he saw about a hundred vehicles coming to a standstill one after another, unable to cross the river and becoming easy prey for bombing raids.

He told himself, "I must reach the other bank at any cost to collect the data I need to design a cable bridge. If we succeed in building such a bridge here, at a spot that is continually targeted by American planes, we will gain valuable experience that we can use elsewhere. At the same time, we will build the confidence of our engineers and workers."

Without hesitation, he threw himself into the violent waters and swam to the other bank, putting his life at risk. The next day, having obtained the measurements he needed, he swam back to his starting point. He set to work for several days, aligning formula after formula and solving equations without a computer.

The construction of the Cà Tang cable was the most difficult of its type. It allowed vehicles to cross even during high floods. Before this, the Vanguard Youth teams had to dive into the cold water to put heavy stone blocks across the ford. It was often a difficult tack because the violent stream would carry away everything and prevent the vehicles from crossing.

As the American war escalated, he and his colleagues continued to perfect their cable bridge models. Some of them allowed 10-ton trucks to cross river over 200 meters wide. After peace was restored in 1975, he used his experience with cable bridges as a basis for a report on the theory of "stability on eslatic cables" in an international conference on mechanics held in Warsaw.

Nguyễn Văn Hường was nominated to the post of interdisciplinary professor of construction and hydraulics and Vice Director of the Council of Ministers (in charge of science and education). He was also elected a member of the National Assembly. He died in 1996 at the age of sixty-nine. His death was mourned by many.

ĐẶNG VŨ KHÚC – AN ENTHUSIASTIC PALEONTOLOGIST

I had spent more than half of my life in Hà Nội without ever setting foot in the Geologist Museum, located behind the Opera House. It was an eye-opening experience for an outsider like me to be inside the walls of this shrine to science. My discovery of the museum was also tinged with pride.

In the summer of 1898, in the middle of the French colonial war, the Governor General created the Geological Service of Indochina. From then until Việt Nam regained its independence in 1945, not a single Vietnamese received a specialized degree in geology. And yet, paging through the bibliography of geological research in Việt Nam over the last hundred years, I was pleased to see many Vietnamese names among the names of French, Russian, British, American, German, Japanese, and Chinese scientists. The internationally-recognized contributions of Vietnamese comprised a good part of the list. One name stood out: paleontologist Đặng Vũ Khúc, with 102 works in all in the bibliography, and who discovered seventy-two new paleontological species over his long career.

The life and education of Đặng Vũ Khúc, who now continues to work even in his retirement, is typical of the second generation of modern scientists trained during the two wars of resistance after the 1945 Revolution. Born into a family of patriotic Confucian scholars in Đa Tốn Village on Gia Lâm District, outside Hà Nội, he left home in 1948 at age seventeen to join the fight against French colonialism. He traversed hundreds of kilometers on foot in order to study in the schools of the liberated zone, including the technical school of Military Zone 4 in Nghệ An and the Transportation and Communications School in Tuyên Quang. Along the way, he still found time to translate Alphonse Daudet's *Lettres de mon Moulin* into Vietnamese and read the novel *Les Etoiles* with his classmate, translating as they went along.

During the closing years of the war, he participated in the construction of Gia Phù-Xồm Lồm Road used during the Điện Biên Phủ Operation in 1953-54, then Sam Neura-Bản Ban Road to assist the Pathet Lao allies consolidate their positions. With the liberation of Hà Nội in October 1954, he returned to the capital to continue his studies at the Polytechnic University. He successfully completed a course in paleontology, one of the first people to do so in Việt Nam. When I visited the Geological Museum, I had the honor of meeting Đặng Vũ Khúc. He led me to a grayish-silver piece of stone in the shape of a mussel. It was labeled *Songdaella graciosa Vukhuc*. This bivalve mollusk fossil dates from 210 million years ago during the Triassic, the richest period of the second geological era.

"*Vukhuc* refers to your name. But what do *songdaellla* and *graciosa* mean?" I asked.

The professor responded, "*Songdaella* is a Latinized form of the geographic name *Sông Đà* (the Black River), where I found the fossil. The Black River haunts my memories. I went and down many times during my youth in the resistance. Once, later on, I spent two days on the river in a sampan that was made specifically for our geological work. The ethnic Thái living along the river were astounded to see it, since they thought sampans could never pass upstream through the rapids. As for *graciosa*, that comes from the Latin word *graciosus* meaning 'graceful.' I wanted to draw attention to the graceful form of these fossils."

I noticed that the inside of the valves did indeed have very graceful concentric curves and the edge of the shell was serrated which allowed the valves to close hermetically, protecting the soft body of the mollusk from the swirling waves of the ocean. The fossil was a true marvel of nature.

Đặng Vũ Khúc informed me that he had discovered three additional species belonging to the genus *Songdaella*. *Songdaella elliptica Vukhuc* was named for its curved shape, and *Songdaella choboensis Vukhuc* in honor of the small market town of Chợ Bờ along the banks of the Black River. And *Songdaella* is just one of the eight genii discovered by Đặng Vũ Khúc. If one counts species, he has discovered a total of sixty-five Triassic bivalves (*Bibalvia*) and nine ammonties (*Ammoidea*) from the Triassic and Jurassic, the first two periods during which dinosaurs grazed on the earth.

Vietnamese paleontologists have made great contributions. In 1995, on the fiftieth anniversary of the 1945 Revolution, Tống Duy Thanh and Đặng Vũ Khúc (president and vice-president of the Paleobiological and Geological Society of Việt Nam) counted 267 new species discovered by Vietnamese scientists, compared to the thirty-five discoveries during the half-century of existence of the Geological Service of (French) Indochina. On the recommendation of Soviet and Vietnamese experts (A.E. Dovzhikov and Nguyễn Văn Chiển), these discoveries were always carefully controlled and named so as to be recognized at an international level. In particular, in research on mineral resources, it was important to determine exactly at which geologic stratum and minerals were found. As for Đặng Vũ Khúc, he discovered eleven new sedimentary strata used in the formation of a geologic map of Việt Nam. This achievement has been usefully theoretically and practically.

The Vietnamese geologists trained after the 1945 Revolution have contributed, along with foreign scientific colleagues, to producing detailed geologic maps of Việt Nam, Laos and Cambodia at a scale of 1/500,000 and 1/1,000,000. After national reunification in 1975, Vietnamese geologists, without foreign assistance, worked on creating a geologic map of southern Việt Nam at a scale of 1/500,000. Đặng Vũ Khúc was responsible for the portions of this survey dealing with paleobiology. He also published studies in English on the paleobiology and paleogeography of Việt Nam.

In October 1978, he defended his Ph.D. thesis at the School of Mines in Leningard (St. Petersburg). It was entitled "The Fauna and Carboniferous Sedimentary Layers in the Early Triassic in Northwestern Việt Nam." Twelve years later, at the same institution, he completed his doctorate of sciences thesis titled "The Triassic Period in Việt Nam: Stratigraphic Paleobiology and Paleogeography."

One telling anecdote comes from this period. In 1989, while in Leningrad, he received a telegram from his colleague Tống Duy Thanh, who was then in Paris. The French Museum of Natural History had lost the labels of several specimens that had been collected along ago by French scientists. They were having difficulty identifying the sample and asked the two Vietnamese scientists for help. He quickly was able to resolve the enigma.

From 1991 to 1995, he was the co-leader of two projects among the twenty projects of UNESCO's international Geological Correlation Program (IGCP). Together with French scientist Henry Fontaine, he carried responsibility for the project studying stratigraphic correlation in Southeast Asia. The seminar on this subject held in November 1995 brought eighty foreign geologists from seventeen countries to Hà Nội, among them was French academician Pierre Taquet, who had discovered an entire fossilized dinosaur skeleton in Laos. But the foreign scientists were outnumbered by far by the 120 Vietnamese geologists who attended the seminar – a sign of successful development in which Đặng Vũ Khúc played an important part.

ĐÀO THẾ TUẤN AND THE "GREEN REVOLUTION" IN VIỆT NAM

Professor Đào Thế Tuấn was born in 1931 in a famous intellectual family in Huế. His father, scholar Đào Duy Anh, compiled two great, highly-appreciated dictionaries. They include *Chinese-Vietnamese Dictionary* published for the first time in 1931 and *French-Vietnamese Dictionary* published in 1936. The work of compiling dictionaries, which normally requires the effort of a research institute or a collective body of multisectoral experts, was carried out by him alone. Many intellectuals must have remembered that it was the strong-willed patriotic scholar Phan Bội Châu (1867-1940), under the pen-name Hãn Mạn Tử, who wrote the epigraph for the great work entitled *Chinese-Vietnamese Dictionary*. In *Larousse Encyclopedia* (France), the word "Đào Duy Anh" was described as "a great name among contemporary encyclopedists."

Along with scholar Huỳnh Thúc Kháng,[1] scholar Đào Duy Anh founded *Tiếng Dân* (The People's Voice) newspaper in 1927. It was the patriotic and progressive newspaper at

1. Huỳnh Thúc Kháng (1876-1947) was a strong-willed patriotic scholar. After the August Revolution (1945), he became Minister of Interior and later Vietnamese Acting President in 1946.

that time. "Quan Hải tùng thư" was a bookstore opened by Đào Duy Anh, Phan Đăng Lưu, Võ Liêm Sơn, and Trần Đình Nam. It was the first place to have progressive printed social science books in Vietnamese script. Đào Duy Anh wrote many significant works such as *Việt Nam văn hóa sử cương* (General History of Vietnamese Culture) in 1938, *Khảo luận về Kim Vân Kiều* (A Treatise on Kim Van Kieu) in 1943, *Khổng giáo phê bình tiểu luận* (Essay on Criticism Against Confucianism) in 1943, and *Trung Hoa sử cương* (General Chinese History) in 1944.

Born into a great intellectual family with solemn and respectful family traditions, Đào Thế Tuấn began learning formally and effectively at a young age. His father used to say to him, "If you don't learn Chinese, it will be difficult for you to study social sciences well." Therefore, when he was young, he was taught Chinese by a Confucian scholar whose job was to record Chinese books as materials. Thanks to such basic Chinese knowledge, when taking part in the pre-intermediate and intermediate officer training course for ground troops in Kunming (China), he could learn Mandarin Chinese.

During the successful August Revolution in 1945, he was learning at Huế National Secondary School where French was being taught. This would help him in his later study. At the same time, he learned English as a foreign language.

He was enthusiastic in joining the "throw aside the writing brush and join the army" campaign, taking part in the Communication Team of Huế Liberation Army, and then moving to the Voluntary Propaganda Team of Việt Minh in the central part of Việt Nam. In 1950, he went to Nguyễn Thượng Hiền High School in the free land of

Thanh Hóa Province (Central Việt Nam) to take the junior bachelor examination. During this time there were soldier recruitments. He volunteered to join the army. After working in the General Staff Department, he was sent to China to attend a one-year training course for pre-intermediate and intermediate officers. He then returned to Việt Nam and took part in the Tây Bắc (Northwest) campaign. In 1953, when he was about to leave for Điện Biên Phủ, he was sent to the Soviet Union for study.

Receiving a Doctor Degree Right After Finishing University Education

Đào Thế Tuấn was very eager to be one of the first Vietnamese students to go to study in the land of the October Revolution. As he recalled,

"I was sent to the University of Tashkent in the capital of the Soviet Republic of Uzbekistan to study about agriculture. There I made friends with some Chinese students and tutored them with Russian. Through that I learned Chinese. Thus, I had no trouble when reading Chinese documents about economic reform and rural development. China and Việt Nam are socialistic countries with a lot of similarities. However, due to historic situations, China often experienced things ten years prior to Việt Nam. Such experiences are useful lessons if we know how to learn....

When I was a university student, I myself did research and wrote a 100-page book called *Nguồn gốc, phân loại và sinh thái cây* lúa (Origin, Classification and Ecology of the Rice Plant), which was published by Uzbekistan National Publishing House in Tashkent. I also wrote three articles on the rice plant: *Xuân hóa lúa* (Springizing Rice), *Lại bàn về xuân hóa lúa* (More About the Springization of Rice, and *Nguồn gốc cây lúa ở Trung Á* (The Root of Rice in Central Asia), which were made public through scientific journals in the Soviet Republic of Uzbekistan. As a result, when I finished my university education, my graduation thesis was recognised to meet the criteria of a doctor thesis by the university's science council. Thus, without going through the three years of postgraduate education, I was one of the students who received the Ph.D. degree in the Soviet Union."

He came back to Việt Nam when he was twenty-seven years old and worked for the Việt Nam Academy of Agriculture and Forestry and then the Việt Nam Academy of Agricultural Sciences. Since then, his studies have aimed at building scientific foundations for the development of agriculture and rural areas in the Red River delta (North Việt Nam). Most of those studies are about plant physiology, and many of them are about rural economics and society.

Devoting His Heart and Mind to "Less High-in-Rank" Research

From 1958 to 1970, he did research on subjects that were considered to be "not high-ranking." This included research on the falling rice phenomenon, which was the basis for selecting short and fertile rice varieties. He then studied the physiological basis of rice by using fertilizer. Through this he could explain the reason why plants in Việt Nam need phosphorus for growth even though using phosphorus is unproductive. Thus, he suggested "turning phosphorus into nitrogen" by supplying phosphorus to azollas and sesbanias and then using them as green fertilizer to manure rice. Such results were respectfully reported in the Asian-African Science Conference in Beijing in 1963.

In addition, he conducted research on the phenomenon of rice plants having only husky rice grains and found a solution that would prevent rice plants from withering away. Rice plants withering away had resulted in many heavy losses in Southeast Asian countries. ,As a result, the disease was stamped out in a short time, and two new rice varieties that could bear the diseases were selected.

Moreover, he summarized all the experiences in cultivating high-yield rice fields. Thanks to it, he built a scientific foundation on intensive rice cultivation that yielded ten tonnes of rice per hectare in the 1970s. In his book *Sinh lý ruộng lúa năng suất cao* (High-Yield Rice Physiology), which was published by the Science-Technology Publishing House in Hà Nội in 1970, he sconveyed the method of getting access to rice population physiology instead of individual access of the rice plants. As a result, a new research method came into being.

In the years after that, he and his assistants studied the physiological foundations for selecting high-yield rice varieties and put forward a selection method based on origin and capacity. At the same time, they did research on the physiological foundation of the selection of rice varieties that could endure extreme ecological conditions such as acidity, salinity, coldness, dryness, flood, and insects. They realized that high-productive rice plants must be short, straight-leafed with a high photosynthetic performance in their last period, and big-flowered. He also helped rice seed selectors create a series of high-yield rice varieties in Việt Nam that met the requirement for intensive rice cultivation in different ecological areas. Along with his assistants, he created some rice varieties such as NN75-10, CN2, V14, V15, V18, 79-1, and CR203, all of which became very popular in rice production. They were the first high-yield rice varieties in Việt Nam.

He recognised that it was not enough for people to only pay attention to increasing productivity while not improving the structure of the plants. From this fact he wrote the book *Bố trí cây trồng hợp lý ở hợp tác xã* (Working Out a Reasonable Structure of Plants in Cooperatives) published in 1969 by Countryside Publishing House in Hà Nội. This book laid the foundation for increasing crops and developing winter crops in Việt Nam. In addition, he did research on bringing temperate plants such as potatoes, wheat, barley, and some other vegetables to Việt Nam. After studying the law of winter in Việt Nam, he suggested dividing winter plants into two types: hot-bearing plants, which need sowing early, and cold-bearing plants, which need sowing late. Then, he developed corn and soy-bean in winter. These studies contributed to widening areas of winter-crops in Việt Nam, making winter crops the main crops and improving farmers' income.

The studies on rice plant physiology and cultivation systems were initial ones in Việt Nam. They paved a new way for the green revolution in Việt Nam and were applied widely by internal institutes. The results of such studies were highly appreciated by the International Rice Research Institute and the Lenin Academy of Agricultural Sciences (formerly in the Soviet Union and presently in the Russian Federation). In 1980, he was nominated professor.

Sensitivity to Modern Agronomy Schools

Having discovered new things, in 1988, he established the Agrarian Systems Department in order to examine a systematic view of agronomy and to build up foundations for comprehensive rural development. He also applied modern views of research in Việt Nam on topics such as household economies, institutional economics, and value chain economics. Along with this, he conducted new ways of doing research for the social sciences in Việt Nam. His works were well-known in France.

His book *Kinh tế hộ gia đình* (Household Economy) was printed by the National Politics Publishing House in Hà Nội in 1997. In his view, the household economy plays an important role not only during the "budget subsidies period" but also at present and in the future.

According to him, the correct way to improve Vietnamese rural areas was not by asking non-agricultural businesses to "step in," for example, investing in agriculture, buying cultivated land of farmers, opening large farms, and hiring landless farmers like the methods under capitalism. Instead, the middle peasants with skills and experiences in farming and wide knowledge of their habitats should be helped to widen their household farms and do cultivation on their own. At the same time, it was necessary to create new work for poor peasants who no longer participated in agricultural activities as well as help them either move to urban areas legally or stay in their countryside to make services or private industries.

Being Honored Both at Home and Abroad

Professor Đào Thế Tuấn is a great writer. In Confucian terms, he is a scholar whose "books are as high as his head."

When he was a university student, he had his first book about rice plants published in Tashkent, Uzbekistan. When he returned to Việt Nam, his other eighteen books were published one after the other by leading publishers.

He also made public hundreds of updated scientific articles. By the time he was presented with the Hồ Chí Minh Award in 2005, he had published 206 works in Vietnamese, thirty-six in English, fifty-eight in French, and five in Russian. What a respectable writer!

Apart from works on plant physiology, he also wrote works about social sciences. They included *Việt Nam có phải là quê hương của cây lúa không?* (Is Việt Nam the Native Land of Rice Plants?), *Về nguồn gốc cây lúa ở Trung Quốc* (On the Rice's Origin in China), *Làng ở vùng châu thổ sông Hồng - vấn đề còn bỏ* ngỏ (Villages in the Red River Delta – Unsolved Issue), *Từ Mộ Trạch đến lịch sử nông nghiệp đồng bằng sông Hồng* (From Mộ Trạch to the History of Agriculture in the Red River Delta), *Dân số Việt Nam trong lịch sử* (Population of Việt Nam in the Past), *Người nông dân châu thổ Bắc Kỳ* (Farmers in the Northern Delta), *Nguyễn Ái Quốc với nông dân* (Nguyễn Ái Quốc and Farmers), *Kinh tế hộ nông dân và thể chế tập thể ở Việt Nam* (Family Farming Economy and Collective Institutions in Việt Nam), *Hệ thống nông nghiệp và vấn đề nghiên cứu xã hội học ở Việt Nam* (Agrarian Systems and Sociological Research in Việt Nam), and *Văn hóa và phát triển* (Culture and Development).

His works on natural and social sciences were published in magazines and journals at home and overseas. They were also presented in international conferences about Vietnamese studies, such as conferences in the Far Eastern School in France and science conferences in Russia, China,

France, the USA, England, Holland, Switzerland, Canada, Brazil, Cuba, Thailand, and other countries.

He was honored not only at home with the title of Labour Hero and the Hồ Chí Minh Award but also abroad. In 1985, the Lenin Academy of Agricultural Sciences gave him the title of Foreign Member. The French government presented him the medal of Knight of Agricultural Merit (third class) in 2001, then the medal of Academic Palm Officer (second class), and recently the medal of Commander of Agricultural Merit (first class). French agronomists called him "the father of the green revolution in Việt Nam."

After retiring, he set up the Association of Science and Rural Development and was elected as the chairman. Before the foundation of the association, people had only focused on increasing the agricultural productivity but paid no attention to the sustainable livelihoods of farmers in terms of rural development. Therefore, the association's activities would be very important. According to him, later rural development programs should be more collective and should pay attention not only to agricultural development but also to the development of an all-sided rural economy with the participation of the public.

Along with the Việt Nam Ministry of Agriculture and Rural Development, the association would work out scientific contents and strategies for the collective rural development in the following decades.

Being proficient in many foreign languages and having knowledge of basic sciences, Professor Đào Thế Tuấn spent a lot of time pondering over modern research works about rural development in the world. He explored the different ideas of economists such as Tchayanov, A. Lewis, Johnston, Mellor, T. Shultz, P. Gourou, R. Dumont, M. Masoyer, F. Ellis, R. Putman, and M. Porter.

For more than three decades since the unification of Việt Nam, as the leader of the Việt Nam Institute of Agricultural Sciences, he collaborated closely and continuously with many famous French agronomists in the Red River program. He also carried out Việt Nam-France co-operative agricultural projects in the Red River delta. He fostered good relationships with many French agrarian professors. He had chance to get access to modern knowledge of agronomy. His son, Đào Thế Anh, successfully defended his doctorate thesis in agronomy at the University of Montpellier, France, and currently is the director of the Center for Agrarian Systems Research and Development of the Việt Nam Institute of Agricultural Sciences. He will be a great successor and will follow his father's footsteps.

With a deep and comprehensive knowledge and experiences, Professor Đào Thế Tuấn applied new solutions to Việt Nam's situation. When proposing a strategy for collective rural development, he hoped to eradicate poverty and stimulate stable development in the Vietnamese countryside.

A Happy and Active Life at the Age of Eighty

I paid a visit to Professor Đào Thế Tuấn on a fine mid-autumn's day. His house is far from the center. It is located on Trần Duy Hưng Street, a new street of Hà Nội. He "showed off" a new book named *À la Découverte des Villages de Métier au Vietnam, Dix Itinéraires autour Hanoi* (The Discovery of Craft Villages, Ten Itineraries Around Hà Nội), which was written by his two French students. On the first page of the book, the authors, Sylvie Fanchette and Nicholas Stedman, had written a dedication to Professor Đào Thế Tuấn. It was him who had given suggestions and directions to Sylvie, a geology doctor, and Nicholas, a master in literature, in writing this book. It was published by Thế Giới Publishers in mid September 2009.

While we were talking with each other, he offered me a strange moon cake with thick green tea stuffing that was acrid and sweet tasting. Then he asked me, "You are wondering why there are so many types of teapots in my house, aren't you? I'm thinking about how to make the Vietnamese tea ceremony become special.... Nowadays, Việt Nam is recognized as the 'cradle of tea' by scientists in the world. Drinking tea is often associated with sitting in meditation and keeping your mind at peace. We need to highlight a special feature of the Vietnamese tea ceremony, don't we?"

I took a meditative look at the man I was talking to. In a few months, he would reach the age of eighty. Then I wondered what made a man who was suffering from diabetes and gastric and liverish diseases so enthusiastic as if he was still a young student in Tashkent. There is no elixir of life. However, a man's passion for his career to live happily and usefully can last until the end of life.

It reminded me of his speech in French entitled *Ma vie lie aux paysans* (My Life is Bound to the Peasants). He made it in a seminar in 2003 where he was given the René Dumont Award by French agronomists. The speech's title is the best description for his life.

He said, "During the destructive war, I was appointed to work for three years in a countryside in Nghệ An Province.

There I ate, lived, and worked together with the farmers. I guided several junior agricultural engineers in assisting farmers to run co-operatives under the US bomb attacks. Those three difficult and dangerous years gave me a profound understanding about the greatest aspiration of farmers to their cultivated land. I experienced the feelings of anxiety and torment over ineffective solutions. This pushed me to devote all my mind and heart in doing research on subjects that are practical in Vietnamese situations.

René Dumont was a French scholar who had done world-known research on agriculture, rural areas, and farmers in Asian and African countries. He always gave support to poor farmers. After his death, French agronomists set up the René Dumont Fund to award the outstanding agronomists who associate their life and career with farmers all over the world. It's my honor to receive the award under the name of this outstanding and noble scholar."

NGUYỄN TÀI THU WITH MIRACULOUS NEEDLES

An article in the French newspaper *La Dépêche*, "Miraculous Needles Make the Mute Speak," describes the story of a mute child treated for months by costly modern methods without result in a hospital in France. Finally, the child visited Vietnamese acupuncturist Nguyễn Tài Thu and could speak within less than ten minutes.

In a hospital in Metz, France, another case attracted public attention. A young woman of around twenty was hospitalized for twelve hours with severe labor pains but was unable to give birth. The doctors decided to perform a Caesarean section. However, at that moment, Professor Nguyễn Tài Thu, who was participating in an acupuncture conference in Paris, arrived. With six needles he relieved the pain of the woman who promptly gave birth to a healthy baby. "Six needle replace a Caesarean," the newspapers proclaimed.

However, it was Professor Nguyễn Tài Thu's practice of acupunctural anesthesia that brought him the most attention. In a gallstone operation supervised by Dr. Bismuth, Professor Nguyễn Tài Thu rapidly inserted an acupuncture needle in the patient's skin so as not to cause any pain. If the needle were placed in the incorrect spot, or inserted too deeply, the patient would feel extreme pain that would increase when the needle was removed or when electropuncture was later applied. Twenty minutes later, he pinched the patient's stomach with force. She did not react. Thus, with the patient was still conscious but numb in the area of the operation, Dr. Bismuth made a precise incision in her stomach, all the while carrying on a conversation with her. The liver and gall bladder was extracted from the abdominal cavity, and the surgeon removed three gallstones. During this more than hour-long operation, the patient felt no pain but a slight discomfort.

"Does it hurt?" asked Dr. Bismuth.

"No," the patient responded. "Can I have a cup of coffee, please?"

The nurse raised the patient's head so that she could drink the coffee.

"Thank you very much," she smiled with confidence.

The Vietnamese acupuncturist was given the title of professor and offered the robe of Rabelais by the Faculty of Medicine in Montpellier, France, the oldest medical school in Europe. He was also made the honorary president of the French Acupuncture Association. The French prime minister and the mayor of Lille gave a banquet in his honor. He was also elected as a member of the French Academy of Acupuncture and made an honorary professor at several other French universities.

From October 17-20, 1991, Columbia University in New York held the Seventh Acupuncture Conference in the

United States with the participation of acupuncturists from fourteen countries around the world with strong traditions in the field. He was invited to chair the first session. After a short opening speech by Japanese-American doctor Yoshiaki Omura, the participants discussed his reports titled "Diagnosis and Treatment of Pain in the Heart and Tongue in Oriental Medicine." He then continued with a description of acupunctural anesthesia in surgical operations. He also showed video cassettes of some typical cases he treated.

The viewers attentively followed the course of an operation at the Hospital Number One in Rome. The patient suffered from cirrhosis of the liver and was in a state of total hemorrhage. Since conventional anesthesia would have been very dangerous in this state, the doctors decided to resort to acupunctural anesthesia. The operation was televised live on RAI Italian television. Nguyễn Tài Thu showed his art in front of the camera, while the patient ate and drank during the operation.

This operation "astounded all of Italy", according to the RAI commentator. The Italian president asked Nguyễn Tài Thu to care for his daughter. Almost a year later, he invited Nguyễn Tài Thu to visit Italy for ten days as an honored guest who had contributed greatly to the development of acupuncture in the country.

At the conference in New York, Cao Dinhgao, Vice-President of the Chinese Association of Acupuncture, told Nguyễn Tài Thu, "The achievements of Vietnamese acupuncture are remarkable. Please do not forget to invite us to your conferences." These were very powerful words as China has always been known for its acupuncture.

After the conference, Nguyễn Tài Thu continued on to Baltimore, Washington, Los Angeles, and Westminster (California). By this time, he was co-president of the Western Pacific Acupunctural Council, co-president of the International Association of Acupuncture, vice-president of the Association of Acupuncture of Europe, and an honorary professor in many countries. Still, he remained modest. He acted the same as at the first time I met him in 1964, as if he was still a novice.

He told me, "If I can achieve anything, it is above all because I follow the example of the ancients. According to ancient literature, acupuncture was practiced with success during the era of the Hùng Kings, 1000 years before Christ. The American Acupuncture Association contains 6,000 members, while in Việt Nam we have 20,000, including many talented young people. Thanks in part to their achievements, I have become well-known by my foreign colleagues."

He added, "But acupuncture is not a miracle cure for disease. Even acupunctural anesthesia only succeeds ninety-seven percent of the time. There are cases where, in the course of the operation, one has to accept painful failures or mistakes. Besides, in my private life, as you know, fate has not spared me of misfortunes..."

The life and career of Nguyễn Tài Thu is closely linked to the August Revolution of 1945 and the two wars of national resistance. As a member of Youth Brigades for National Salvation in 1945, he became the head of a propaganda team in Hoàn Kiếm District (Hà Nội) at the beginning of resistance in 1945. At the age of seventeen, he left the capital for missions in the mountainous provinces of Lào Cai and Yên Bái. After graduating from Hùng Vương Secondary

School in Phú Thọ, he entered the army. As a soldier in Engineering Regiment 51, he was admitted into the Medical and Pharmaceutical College where his professors were eminent doctors such as Hồ Đắc Di, Tôn Thất Tùng, Đặng Văn Ngữ, and Đặng Vũ Hỷ.

At the Việt Nam Institute of Acupuncture, created and directed by him, most of the patients are disabled or disadvantaged. Many are war veterans who have suffered for their country and who now, in their old age, possess nothing but the most serious diseases: cerebral hemorrhages, total or partial paralysis, rickets, blindness, or hearing loss. The children in the hospital are also disadvantaged, afflicted with paralysis or the consequences of encephalitis, cerebral hemorrhages, or other birth defects. Others have visual difficulties, deafness, or muteness, either congenital or acquired.

Since its foundation, the Institute of Acupuncture has never required poor patients to pay for their hospitalization. Several years ago, the Ministry of Health standardized hospital fees, but the doctors at the institute did not have the heart to carry this through. On the contrary, the institute provides food for poor children and gives them presents on children's holidays (International Children's Day on June 1 and the Mid-Autumn Festival). All patients who spend the Lunar New Year's festival at the institute receive traditional cakes, snacks, and a little money.

The Institute of Acupuncture is located on Si Field in the village of Vĩnh Hồ, now part of Đống Đa District (Hà Nội). When I visited there, I met Dr. Bùi Thị Hiếu. She said, "The institute was founded in April 1982 by a decision of the Minister of Health, but it was given no site. For an entire year, Nguyễn Tài Thu had to bicycle through Hà Nội and the surrounding villages looking for a place to build the institute. Finally he found land on Si Field and insisted that the People's Committee of the City reserve it for us. At that time, there were marshes here and uncomfortable brick buildings for which they asked us to pay nearly 100,000 dollars. Nguyễn Tài Thu had to pay this sum with his own money, saved from the courses and treatment he had done overseas. But it was only a year after the payment that we were able to set our foot here."

Professor Nguyễn Tài Thu and his colleagues filled in the marshes, dug a pond, planted lotus and matching trees, and laid out paths. During his international travels, he made appeals to various international organizations and Vietnamese nationals to raise funds to purchase four wood houses, which he had transported to Hà Nội to use as treatment rooms. These Finnish-style houses still remain as mementos of the first, difficult period of the institute, next door to the larger buildings that have been constructed recently.

What resources does the Institute of Acupuncture have to construct its facilities, purchase equipment, and come to the assistance of patients?

The answer is simple. Foreign patients from many different countries - French, Italian, Russian, Czech, Australian, Japanese, and Indian - pay an average of ten dollars per treatment. Thanks to the international prestige of Professor Nguyễn Tài Thu, the institute also runs a comfortable mini-hotel and rents these rooms for $20 per night. In addition, it produces needles and electropuncture equipment for the domestic market and for export. Profit from these activities is consigned to the aid of poor patients.

Professor Nguyễn Tài Thu also adds the cash honoraria he receives from courses given abroad on behalf of the institute. In this way, he also practices the philosophy of traditional medicine.

Professor Nguyễn Tài Thu is a worthy successor of the most famous Vietnamese doctors of the past, Lãn Ông and Tuệ Tĩnh.

HOÀNG XUÂN SÍNH – THE FIRST FEMALE PROFESSOR OF MATHEMATICS IN VIỆT NAM

In the medieval patriarchal society in both the Occident and Orient, women were not allowed to hold their deserved ranks in examinations and socio-political activities no matter how erudite and talented they were. Within the nearly 1000 years of Confucian examinations in Việt Nam (1075-1919), nearly 3000 doctorate laureates were recognized. All of them were male. Nguyễn Thị Duệ was the only woman who disguised herself as a man to take the exam. She got a doctoral degree but soon was deprived of all the achievements after her gender was revealed. Hence, her name was not carved on the stone steles in the Temple of Literature (Hà Nội).

Nonetheless, the victory of the August Revolution in 1945 and the birth of the Democratic Republic transformed "women's lots." Since then, many women have been granted as doctors, doctors of sciences, associate professors, or professors without employing "male disguises." Female Professor and doctor of mathematics Hoàng Xuân Sính is a typical example. And she has been followed by some others, such as Lê Hồng Vân and Nguyễn Thị Thiều Hoa, in the "difficult, hard, and dry" mathematical field.

Attaching Herself to the National Destiny

Hoàng Xuân Sính returned to Paris where she had spent her student life and had gained the degrees of Bachelor of Sciences and then Master of Mathematics. There she secretly joined the movement of patriotic overseas Vietnamese led by pioneer intellectuals such as Phạm Huy Thông and Nguyễn Khắc Viện. Following these examples, she voluntarily left the peaceful and convenient Western country for her hometown where the anti-American war was in its fiercest phase. However, her boyfriend insisted on staying behind, so they both agreed to break up. President Hồ Chí Minh's lofty example of sacrifice began stimulating her patriotism when she was still at high school in Hà Nội. During that time, the city was occupied by the French expeditionary force (1947-1954). She bravely joined the secret anti-colonialism movement of patriotic students.

On her return to Việt Nam, she was assigned to lecture mathematics at Hà Nội National University of Education. Nonetheless, like many other Vietnamese mathematicians, she had to evacuate from Hà Nội to the countryside, a place isolated from the science world. At that time, modern technologies such as laptops, internet, emails, and websites were not available yet. Mathematical journals from foreign countries, especially from developed countries such as America, England, France, the Federal Republic of Germany, and Japan were few in number in Việt Nam and usually came quite late to Việt Nam. Consequently, she could neither get access to her fellow international mathematicians' latest works nor grasp great ideas which would suggest to her some promising topics. She had wished to do international-level research on such topics so that they could be published in first-rate mathematical journals.

Regardless of the escalation in the American bombing of North Việt Nam, quite a few famous foreign mathematicians, some of whom had even won the Fields Medal (regarded as the Nobel Prize in mathematics) supported the struggle of Viet Nam and came to Hà Nội. They gave lectures on modern mathematics in seminars organized in mountainous Đại Từ (Thái Nguyên Province). Many big contemporary universities had evacuated there. Among these people, Hoàng Xuân Sính especially admired French Jewish Professor Alexander Grothendieck, an authority on algebraic geometry and an owner of the 1966 Fields Medal. This 39-year-old professor's lectures gave her an overview about key matters of modern mathematics and helped direct her research in the right direction. However, doing a several-hundred-page State doctorate (docteur d'État) thesis also required many other difficult steps. Mathematics was often assumed to be a "very difficult, hard, and dry" field. These first two features of mathematics were rightly inevitable. However, the term "dry" depended on personal tastes. As for her, a valid theorem was as captivating as a meaningful poem.

Her State doctorate thesis was written in the dim oil-lamp light in a poorly-furnished cottage in a village by the Đáy River in North Việt Nam. During the years of pursuing the thesis, she could not consult her supervisor but her colleagues. Regardless of persistent rains, chilly northeast monsoons, multi-hole wooden walls, and flickering oil lamp light, she still huddled herself up in an old thin grey blanket to write down her new ideas.

Defending the State Doctorate Thesis in Paris – "the city of light"

Allowed by the Vietnamese government, Hoàng Xuân Sính brought her State doctorate thesis to Paris to defend it. Setting foot in noisy and gaudy cities like Paris, she always felt lost. This feeling had persisted during her student life in France. This return was not an exception as she did not like the so-called luxury, noise, and business. She, with an "Oriental mind," seemed to be only suitable in an environment of simplicity, peace, and deliberation all of which made her deep in thought. However, owing to this return, she had a chance to meet her old comrades in the anti-French movement of patriotic overseas Vietnamese. Moreover, she could visit the mathematician and Professor Grothendieck, who had given her lectures on algebraic geometry in Việt Nam and had been nearly hit by bombs there.

The defense of her thesis was made at Université Paris Diderot (Paris Diderot University). The thesis evaluation committee consisted of many famous mathematicians, including Professor Henri Cartan, an academician of the French Academy of Sciences, and Professor Alexander Grothendieck. Both of them were awarded Fields Medals. Her defense of the first thesis *Gr - catégories strictes* (Gr - Strict Categories) written during years of evacuation in a village by the Đáy River, lasted two hours and a half. Shortly after that, she had to uphold the second one *Immersion d'un complexe à une dimension dans une variété différentielle à deux dimensions* (Immersion of a One-Dimensional Complex Into a Two-Dimensional Differential Variety), which was done within two months in Paris. Its topic was given by the Council of Mathematics of Université Paris Diderot to "test her talent." Both of the theses were aimed at solving issues of modern mathematics in different angles.

Hoàng Xuân Sính was the first female foreigner to succeed in defending her State doctorate thesis in mathematics in Paris.

Information about her was honorably introduced in the August 1975 issue of the monthly magazine *Soviet Women* published in Moscow. The introduction took up half of the page on "Vietnamese women." In an interview with a Soviet journalist, she spoke out her dream, "I wish to help my country produce a generation of young mathematicians. Now many post-graduates still have to go abroad to work on their doctoral theses. We hope to set up many high-level majors of mathematics in Việt Nam".

Springer-Verlag Publisher, a world-famous publisher of scientific books with branches in Germany, England, and America, also asked her to convert the thesis into a monograph for publication.

Before defending her thesis in Paris, she had presented her research results at the Việt Nam Congress of Mathematicians in 1971 in Hà Nội and at the International Congress of Mathematicians in 1974 in Vancouver (Canada).

Honoring the Preeminent Russian Female Scientist

Professor Hoàng Xuân Sính was the main contributor to creating the Kovalevskaya Award for excellent female scientists from Việt Nam.

In 1985, after getting royalties from his mathematical books and her book *Sofia Kovalevskaya – Her Life and*

Works, the American couple Mr. Niel and Mrs. Ann Koblitz came up with the idea of creating a scientific fund named Kovalevskaya Fund to encourage women in pursuing their lines of research. The first targets of the fund were women from Việt Nam, Cuba, Nicaragua, El Sanvador, and other places.

Why did Dr. Anna Koblitz name the award created by herself Kovalevskaya Award? Because she realized that Sofia Kovalevskaya was a notable example of scientific interest and excellent research for women not only in Russia but also in other countries in the world to follow.

Recalling the Past Difficulties

Professor Hoàng Xuân Sính recorded, "I still remember that Saturday afternoon in late April 1985. The International Department of the Việt Nam Women's Union called a meeting of some female scientists to its central office on Hàng Chuối Street, Hà Nội, to inform good news about the foundation of the Kovalevskaya Award. The meeting started at 2 p.m., but I was present at 1 p.m. to make some preparations. However, the head of the department sadly told me that our 'superior' (Director N. in the Ministry of Foreign Affairs at that time) did not approve of our proposal because of matters related to the U.S. embargo upon Việt Nam. I wondered how to inform this news to other fellow members and how to explain the matter to progressive American scientists as well.

Cycling home, I was stressed by a dozen of worries. I was cooking dinner for my children, but my mind flew to somewhere else. The following day was Sunday, so I spent the whole day considering what to do. On Monday afternoon, I cycled to the palace of Prime Minister and asked for Permanent Deputy Prime Minister Tố Hữu. Deputy Prime Minister Tố Hữu met me right after that. After grasping the matter, he judged that the proposal was necessary and promised to check and solve it. No sooner had I driven home and just drunk a glass of water than Director N. came. When I, taken by surprise, did not know why he came here, he smiled and said that Deputy Prime Minister Tố Hữu ordered him to ask me more about the story related to the Kovalevskaya Award. After listening to my report, he promised to give a decision within the following morning. In the following afternoon, I met the Koblitzs. They informed me that the authorities concerned met them and welcomed the foundation of the Kovalevskaya Award. I sighed with relief."

Her American fellows offered her the right to select the model for the award medal and certificate. In an American goldsmith's shop she was given some samples, among which "the knowledge lamp" satisfied her the most. The model implied her hope that Vietnamese scientists would be paid proper attention to so that they could work wholeheartedly in the light of knowledge.

For the past two decades, the Kovalevskaya Award has been yearly awarded. Female scientists who have won the award are all happy to have their scientific works publicly known and encouraged. Professor Hoàng Xuân Sính is also pleased as she did a good deed in contributing to the inception of this award.

She is the first Vietnamese woman to be awarded a full professorship in the mathematical field. She has many times been deputed as the chief of the Vietnamese team in the International Mathematical Olympiad.

She has also taken part in a variety of social activities. Everyone who has ever come into contact with her feels that she is quite keen and tactful.

Professor Hoàng Tụy in his private house.

Professor Đào Thế Tuấn chairs a meeting.

Young 26-year-old physicist Nguyễn Văn Hiệu excellently defends his Ph.D. thesis at the Dubna Joint Institute for Nuclear Research, former USSR (1964).

Professor Hà Văn Tấn by his bookcase in his private house.

Professor Nguyễn Tài Cẩn delivers a presentation at the Institute of Literature of the Việt Nam Academy of Social Sciences.

Professor Hoàng Xuân Sính,
first Vietnamese female mathematics professor.

ĐÀO VỌNG ĐỨC AND THE ACADEMY OF SCIENCES OF THE THIRD WORLD

Đào Vọng Đức, born in 1936, was one of my best friends throughout secondary school. We both belonged to the generation of intellectuals formed by the 1945 Revolution and the two wars of national resistance. We were still boys in sixth or seventh grade when the general uprising of August 1945 spread throughout the country, putting an end to the eighty years of French colonization. Assembled into groups of Pioneers for National Salvation and wearing our scarves proudly, we sang independence songs in unison:

"Come quickly, pioneers,
Follow the red flag with a gold star
And answer the demands of our mountains and fields!"

We had just finished elementary school when the war of French re-conquest broke out in 1946. We continued to study in a private secondary school named Tân Dân (New People) in Nam Đàn District, Nghệ An, the home district of Hồ Chí Minh. During the eight years of the war of resistance, Nghệ An was part of the liberated zone. Our classes were taught in Vietnamese, no longer in French. Lacking classical books, we copied our lessons by hand. One student from Professor Nguyễn Thúc Hào's math class lent us Hoàng Xuân Hãn's

Vietnamese Dictionary of Scientific Terms which we recopied by hand over seven days. Using the same method, we even deciphered French math, physics and chemistry textbooks. We also did exercises from the math textbook by François Brachet. The most difficult exercises were assigned to Đào Vọng Đức and me.

In 1952, the two of us were admitted to Huỳnh Thúc Kháng High School through a competitive process. This institution had been evacuated to Bạch Ngọc, a pretty little village on the banks of the Lam River. The main path of the village was bordered with magnificent areca trees, planted by each family when a new child was born. Life was hard in the liberated zone as we were surrounded by the enemy. Students who came from neighboring regions did not have enough money to pay for room and board. Peasant families gave evacuees lodging for free. Similarly, we students organized a "common pot" of a dozen classmates to eat together. Sometimes we were short of rice, our staple food, and had to eat *bù rợ*, a plant that grew abundantly in the area.

None of this prevented us from singing throughout the day with the accompaniment of a banjo, mandolin, or guitar. I still remember the lyrics of one popular Finnish song that we used to sing:

"Here is the wind, the forest and the fields,
Life spreads its board wings.
My darling, why are you so sad?"

We also read each other love poems by Louis Aragon; at our age, we could only dream of love. Later on some afternoons, Đào Vọng Đức and I would sit on the banks of the Lam after having taking a swim. At sunset we looked towards the southeast where the Thiên Nhẫn (Thousand-Foot) mountain range rose on the horizon. That was where the national poet Nguyễn Du (eighteenth century) retired after his accomplished career. Other times we would stretch ourselves out in a cornfield, dreaming of being able to study abroad in Moscow, Berlin, or Prague once the resistance war won in order to become doctors who could thrill the world with our discoveries. Such were our adolescent dreams in the midst of war.

Classes were held in the evening, from seven until ten, for safety from French bombing raids. When it rained, we had to return home on pitch-dark muddy paths, our conical hats and palm-leaf jackets soaked by the heavy downpour. But our hearts were warm as we recited the poem of Nguyễn Đình Thi, "Every step is a step of sacrifice..."

"Splash!" One classmate fell into a ditch full of water, unleashing a cascade of laughter. So unfolded our daily life with its joys and sorrows.

Books were rare, printed on yellowish local paper. The scientific books available to us included *The Life* and *Atoms, Nuclei and Cosmic Rays* by Tạ Quang Bửu, as well as *General Mathematics* and *Theoretical Mechanics* by Nguyễn Xiển, among others. For literature, we had a few novels, essays, and works translated from Russian or Chinese. These were passed from student to student to be devoured.

The dark days of agrarian reform did not shake our faith in the future of the country. After Điện Biên Phủ, the war was over. When we completed our studies, he and I followed divergent paths. I became a journalist while he continued in the path he had chosen. With a diploma in physics from Lomonosov University in Moscow, he defended his doctorate of science thesis at the Dubna Joint Institute for Nuclear Research.

Years later, during a lesson given at the summer session of the International Center for Theoretical Physics in Trieste, Nobel Prize-wining physicist Abdus Salam gave credit to his research. It was Salam, together with academician Nguyễn Văn Hiệu, who engineered his entry into the Academy of Science of the Third World. His doctorate of sciences thesis, *Conformal Invariance of Interactions of Elementary Particles*, treated new problems in quantum theory at the time it was completed in 1974.

When he came back to Việt Nam in 1975, he worked at the Institute of Physics and became director of the Center of Theoretical Physics. Together with Nguyễn Văn Hiệu, he organized yearly conferences to promote research on physics in Việt Nam.

What I admired most about this eminent physicist is not his titles and social standing, but rather his creative spirit and his qualities as a researcher. He avoids the spotlight and likes to work in silence. He never lets a year go by without finishing some interesting article. Seventy of his works have been finished in international physics journals in America, Germany, France, Italy, Holland, Canada, and Japan. He has participated in conferences held in Trieste, Geneva and Dubna. And he has been invited to join the organizing committee of the international conference of the Asia-Pacific Physics Association in Brisbane. But every time we meet, we always go back, not without emotion, to the hard and glorious years of revolution and the war that shaped our generation.

PHAN ĐỨC CHÍNH – THE TEACHER OF YOUNG MATHEMATICAL TALENTS

On the one-hundredth anniversary (1906-2006) of the foundation of Indochina University (present-day Việt Nam National University, Hà Nội), one hundred professors were selected and honored. One of these professors was Associate Professor Dr. Phan Đức Chính, who was selected for his great contribution in training many young mathematical talents for the country.

The Early Days

I came back to the Mathematics and Informatics Gifted Senior High School, the University of Natural Sciences (belonging to Việt Nam National University, Hà Nội) on an ice-cold day in the late winter in 2005. The school had just been awarded the Independence Medal and the title "Hero of Labor during the Renovation Period" by the Vietnamese government. I suddenly recalled the early days of the previous decades when the school was in the deep jungle in Đại Từ, Thái Nguyên Province.

Associate Professor Dr. Phan Đức Chính is a person who has been close to the school since its earliest days.

"Tens of years have passed already. I have taught thousands of mathematically gifted students," said he. "There were so many students from all over the country that I fail to remember them all. However, I still remember and feel proud of two outstandingly successful students. They are Professor and Doctor of Science Đào Trọng Thi, a member of the Central Party Committee and the Director of Việt Nam National University in Hà Nội, and Professor and Doctor of Science Trần Văn Nhung, the Deputy Minister of Education and Training."

Phan Đức Chính graduated from university in 1956 at the age of twenty. He was amongst the scientists who immediately received full-time training after the liberation day of Hà Nội (10 October 1954). He was in the same course with famous Vietnamese physicists such as Nguyễn Văn Hiệu, Vũ Đình Cự, Đàm Trung Bồn, and Vũ Thanh Khiết; historians such as Đinh Xuân Lâm, Phan Huy Lê, Hà Văn Tấn, and Trần Quốc Vượng; and mathematicians like Hoàng Hữu Đường, Nguyễn Thừa Hợp, and Nguyễn Bác Văn.

Phan Đức Chính was born in a family that helped feed and give shelter to revolutionary cadres in Hà Nội when it was temporarily occupied by the French. In 1961, he was sent to the Soviet Union as a research student at the well-known Moscow State University. During his years in Moscow, he spent time writing a Ph.D. thesis and collaborated with Professor G. E. Shylov in compiling a monograph book called *Measure, Integral, Derivative in Linear Space*. This was the first time that a Vietnamese person was a co-author of a book published in the Soviet Union. Returning to Việt Nam in 1965, he taught special subjects such as Stieltjes

Integral, Lie Group, Lie Algebra, Abelian Group, and Galois Group in Faculty of Mathematics at Hà Nội University. These subjects, according to him, are now no longer taught at universities.

Taking Care of Young Talents

Apart from President Hồ Chí Minh, the prime minister at that time, Mr. Phạm Văn Đồng, was greatly devoted to education. He paid much attention to discovering and nurturing young talents, especially those in mathematics.

At that time, there were many specialized classes in secondary schools such as those in mathematics taught by teachers Tôn Thân, Lê Mộng Ngọc, and Vũ Hữu Bình in Trưng Vương School.

Doctor Phan Đức Chính returned to Việt Nam in 1964 when American aircrafts and warships began raiding the northern part of Việt Nam. However, during that trying and dangerous time, Professor Lê Văn Thiêm, the Vice Director of Hà Nội University, and Doctor Hoàng Tụy, the Dean of the Faculty of Mathematics, said to Doctor Phan Đức Chính, "The Prime Minister has just instructed that no matter how fierce the war is our school still must hold mathematics classes for students of senior high school who have high aptitudes. Please lend us a hand."

From then on, along with teaching at university and doing scientific research, he was assigned to teach students in mathematics at specialized senior high schools. This was a time-consuming mission.

Returning to the Forest of Việt Bắc[1]

Professor Phan Đức Chính recorded, "In the early course of 1965-1966, having to cross sections of roads showered with bombs, senior high school students of just around fifty to sixty years of age from different provinces and cities in North Việt Nam gathered in Phúc Xá (Hà Nội). Then they were taken to an evacuated school in Đại Từ (Thái Nguyên Province) – a revolutionary base during the anti-French resistance war. The classrooms were three thatched cottages located in the heart of Đình Hamlet, near the Faculty of Mathematics and the Headquarters of Hà Nội University. Faculties of the university sent teachers of high qualifications and with great enthusiasm to teach the students. The teachers taught and drew experiences from both the curriculum and their own methods. The students studied happily and enthusiastically. In the second course, the mathematics team of the university won nine out of ten prizes in the National Mathematical Olympiad. That was a success beyond expectations!

Along with studying, those students also had to go to the forest, cut down bamboo, and bring it back to the classrooms to help with repairs. They also had to cut and deliver wood to the kitchen. Now, in retrospect, I felt a great pity for them. They had to live far away from home at such a young age."

Professor Đào Trọng Thi, a former student of the second-course Mathematics Specialized Class, said, "We were called back to the school when we were evacuating to the district

1. Việt Bắc was a revolutionary base of the Vietnamese People's Army during the anti-French resistance war (1945-1954). It presently includes six provinces in Northeast Việt Nam: Cao Bằng, Bắc Cạn, Lạng Sơn, Thái Nguyên, Tuyên Quang, and Hà Giang.

of Đại Từ, Thái Nguyên Province. When the city had just lighted up, a wartime train left Hà Nội and took us to the station. The school sent a car to the station to pick us up as a priority. We took turns to walk and sit in the car, going through thirty kilometers of a forest road in the dark. The classroom was set up on a hillock in the middle of a moor surrounded by large dark blue mountains. We went into the forest and cut and brought wood back to the classroom to make simple tables and chairs. Then we dug a tunnel for shelter and a trench across the classroom. Strange as it might seem, we still cherished a dream to become mathematicians even in those extremely hard days!"

While living and studying in the middle of Đại Từ Valley, those students not only improved their knowledge in mathematics but also learned numerous things in the most challenging subject – "becoming useful people."

Taking Students to the International Mathematical Olympiad in Berlin

Early in 1973, the Paris Agreement for restoring peace in Việt Nam was signed. The universities from evacuated areas returned to Hà Nội in succession. In the summer of 1974 when the country was still not reunified, Việt Nam ventured to appoint one team of five students, even though eight were required, to attend the sixteenth International Mathematical Olympiad (IMO) taking place in Berlin, the Democratic Republic of Germany. They were led by teacher Lê Hải Châu and Doctor Phan Đức Chính. Việt Nam took part in the IMO with a faint hope of winning a bronze medal. However, the result turned out incredible. Four out of five students won medals: Hoàng Lê Minh won gold, Vũ Đình Hoà won silver, and Đặng Hoàng Trung and Tạ Hồng Quảng won bronze. Nguyễn Quốc Thắng only needed one more mark to win a bronze medal.

The Astonishing Success of Vietnamese Students

A few days later, the newspaper *Weekly Post-Office* published in Berlin, featured an article titled "Thirty-eight Marks for Hoàng Lê Minh." The article began by mentioning the sixth mathematical paper on polynomials which was considered the most difficult part in the exam by most students. However, it was solved by all five Vietnamese students. The article also mentioned,

"In the competition for solving six papers of various forms of problems, the Vietnamese Delegation has surprised the public. Four out of five students have gained prizes. It is unbelievable that the students of a country which has just recently undergone a cruel destructive war can understand and master the vast and complicated knowledge of mathematics so well.

The Assistant Chief of the Delegation and Mathematics Doctor Phan Đức Chính from Hà Nội University told an anecdote. In 1965, in the middle of a cruel destructive war, President Hồ Chí Minh gave advice to teachers and students in a letter, 'Keep studying and teaching no matter how difficult the situation is.' In accordance with that recommendation, Việt Nam tried to open examinations for mathematically gifted students and set up mathematics specialized classes in high schools.

In as early as December 1972, the National Mathematical Olympiad still took place as usual despite the bomb raids. At that time, Hoàng Lê Minh, who won the first prize in the IMO, was a mathematically gifted nine-grader (equivalent to an eleven-grader now) in a small village in the midland, far away from home...."

Professor Skvortsov, the Chief of the Soviet Union Delegation, said satisfactorily, "A week prior to the exam when Vietnamese students came to Moscow, after several exchanges, we were sure that the Vietnamese team is strong."

Professor Cuculescu, the Chief of the Romanian Delegation, and Professor Tchudanov, the Chief of the Bulgarian Delegation, agreed with Doctor Phan Đức Chính, "It is said that Hoàng Lê Minh is one of your students. Congratulations! You are a teacher of an excellent student."

Continuously Winning Medals

From 1974 to 2005, students at the Mathematic and Informatics Gifted Senior High School won totally fifty-eight medals, including twenty gold medals. In two successive years of attending the IMO (in the eleventh and twelfth grade), Ngô Bảo Châu, Đào Hải Long, Ngô Đắc Tuấn, and Lê Hùng Việt Bảo won eight gold medals.

In 1989, Vietnamese students started to take part in the International Olympiad in Informatics (IOI). From 1989 to 2005, students in the school won twenty-six medals. Nguyễn Ngọc Huy won two gold medals in two successive years by himself.

There are currently over eighty countries taking part in the IMO. Nevertheless, the Vietnamese Delegation is still unofficially ranked as "top ten."

In the report read at the seventh National Congress of Patriotic Competition in October 2005, Doctor Nguyễn Vũ Lương, the dean of the school, said that 400 former students of the school had received Ph.D. degrees and thirty others had become Doctor of Science. Many had become well-known professors and scientists such as Đào Trọng Thi, Trần Văn Nhung, Lê Hồng Vân, Nguyễn Thị Thiều Hoa, Đàm Thanh Sơn, Nguyễn Tiến Dũng, Ngô Bảo Châu, and Ngô Đắc Tuấn. For a high school with a small number of students, this was surely a significant achievement.

Currently, there are not any universities in Việt Nam that have reached the international standard. However, some high schools, such as the Mathematics and Informatics Gifted Senior High School, have reached the international standard.

One day, prior to the Lunar New Year's festival in 2006, I came to visit Associate Professor Phan Đức Chính at his own house on the bank of West Lake. He was very old and had retired. His backbone had been calcified, and his body was pale and slim. However, when he recalled the days of teaching at the Mathematics and Informatics Specialized Senior High School, he was still very passionate and emotional, "I rarely shed tears in my life. Yet, in the summer of 1974 when I took five students of our country to the IMO in Berlin and four of them won prizes, including first and second, I could not prevent tears falling from my eyes."

VŨ ĐÌNH CỰ AND THE GK1 GROUP: CLEARING AMERICAN MINES

After the Điện Biên Phủ Operation, the Geneva Agreement put an end to the French war of reconquest and divided the country into two parts. The war was taken over by the Americans, who investigated the Sài Gòn administration. In 1965, American forces intervened directly in the South, leading a "local war" in an attempt to destroy the people's forces. During this time escalated bombing of the North began.

The Tết Mậu Thân Offensive (1968) caught the Pentagon off guard and shocked American public opinion. From 1969-1972, President Nixon applied the Vietnamization policy, which consisted of supporting the Sài Gòn troops to "fight the last Vietnamese and bring the GIs home from Việt Nam.

However, the new strategy did not have a lasting effect. The people's forces of the South, thanks to reinforcements from the North, launched a large-scale offensive at the beginning of 1972, using heavy tanks and heavy artillery for the first time. Nixon ordered the reinstatement of massive bombing of the North in April. The B-52 Superfortress went into action. Underwater mines were planted to block the principal river mouths of the North, in particular the port of

Hải Phòng. More audaciously than Johnson, who used mines only for tactical purposes, Nixon used strategic sea mines like the MK-52 to block foreign military assistance to the North and deliveries of oil to the port of Hòn Gai. Twenty-six foreign vessels were immobilized. A large number of sailors had to return to their countries by air. The provision of arms, rice, and gasoline to the North was halted.

It was under these circumstances that the GK1 Group was created at the Hà Nội Polytechnic University and charged with clearing the sea mines blocking the port of Hải Phòng. Vũ Đình Cự, one of the few Vietnamese with a doctorate in science at the time, volunteered to head this dangerous mission. He argued that since he was a single bachelor, he was putting only his own life at risk. During a meeting held among the ruins of Hải Phòng, which was being bombed night and day, the chairman of the Department of Maritime Transport of the Ministry of Transportation and Communication told the group, "Our orders are to open a way through no matter what the cost. Our boats along the coast must be able to transport coal from Hòn Gai and Cẩm Phả to other provinces and to bring food and armaments to the port on the Gianh River where trucks will carry it south on Hồ Chí Minh Trail. The navy and the Ministry of Transportation and Communication have already formed special minesweeping units. However, the new mines that Nixon has laid are extremely lethal. We need the help of you, scientists, to clear them."

Two weeks later, Thái Phong, the leader of a ship repair squad who acted as minesweepers, succeeded in clearing a giant MK-52 strategic mine, capable of sinking a 10,000-ton ship. He accomplished this using only the strength of his own muscles, an unprecedented endeavor that could have easily cost him his life since he knew nothing about how to disarm the mine.

Using tools brought from Hà Nội, the members of GK1, clad only in bathing suits, stood around the mine, its cap replaced by a lamp. Why had Nixon proclaimed within seventy-two hours after the mines were laid that boats could leave the territorial waters of Việt Nam without any risk? Why had certain Vietnamese ships coming and going in peace exploded at unforeseen moments? Why in places where minesweepers had passed unharmed did cargo ships later explode? So many problems in Hải Phòng had no solutions to be found. The team had to bring the MK-52 mine back to Hà Nội, where more sophisticated scientific and technical tools were at their disposal.

There, Vũ Đình Cự and his researchers discovered the secret. The mines had delayed fuses set at different times in order to become active at unforeseen moments. At the same time, the team also learned about the use of the new MK-42 mines. While the MK-52 were laid largely along international navigation routes, the smaller MK-42 were spread along the coast, waterways, and railroads. The scope of investigation of the GK1 Group was enlarging.

He passed the "brain" of an MK-42 mine to his assistant, engineer Đào Đức Thành.

"Be careful," he said, "this is the only one of these that we have. If you damage it, it might cost us someone's life to get another!"

How to distinguish the functions and connections of a hundred pieces of metal in a mine as large as a milk crate and as hard as a stone and then reassemble it to test the hypotheses developed in a series of discussions? In his doctoral studies, he researched the physical properties of magnetic membranes. His work had been applied to the development of new computer components. And the command center of the MK-42 mine was itself a sort of computer, consisting of numerous magnetic membranes.

His explanations guided the electronic circuit and current specialists. The group of engineers led by Bùi Minh Tiêu succeeded in outlining a technique of disassembling the mines. Based on this experience, the GK1 Group found a way to jam the MK-42 and MK-52 and prevent their normal functioning. They were assisted in this by the elderly Professor Tạ Quang Bửu, who frequently attended meetings and made many thoughtful suggestions.

Even with the discovery of the special features of the MK-42 and MK-52, the problem was far from solved. In order to clear the mines, it was necessary to create a magnetic field that was comparable to that of an actual ship. How could the researchers estimate the intensity and frequency of magnetic signals given off by ships and the various positions of mines? To assure the correct placement of the de-mining instruments, they relied on the formulas of Russian physicist Lev Landau, Nobel Prize and Lenin Prize winner. Vũ Đình Cự had already used Landau's theories in his work on circular and flattened coil.

Another member of the GK1 Group, Dr. Nguyễn Bính, accepted the task of finding a solution to this problem. He contacted the Computer Center for his assistance, but after a week of work they were unable to write a program to deal with such large amounts of data. Nguyễn Bính realized that he would have to rely on less exact methods of problem solving. Pressed for time, he isolated himself in a farmer's house in Từ Sơn District (Bắc Ninh), a dozen kilometers from Hà Nội, in order to work on the calculations on his own. Wearing shorts and a T-shirt, he spent five days of mind-numbering work looking for a solution. Finally, with the program in his pocket, he returned to the Computer Center in Hà Nội. The Soviet Minsk-22 computer began to process his instructions. He had frequently used computers to solve difficult problems when he was a doctoral student in the Soviet Union. He knew it was a slow, painstaking process. Waiting in the air-conditioned room at the center, he heard convoys of military trucks passing by on their way to the South. He remembered his own time in the army, on the eve of the Điện Biên Phủ battle that would end the French war of reconquest (1946-1954). One day, he was called to meet the political commissar of his battalion and received an order to leave the army to go for many years of technical studies in the USSR. And now here he was, following the efforts of the computer. After working for seven and a haft hours and carrying out 170 million calculations, the computer was finished. The results were satisfactory.

During these days of laborious research, no one in the GK1 Group dreamed of following a regular daily schedule. There were no set times for meals or sleep. Vũ Đình Cự was so absorbed in his work that he would often forget to eat dinner. One evening, when he showed up at the Polytechnic University cafeteria after 8:00 p.m., the head waitress did not even look around to see who he was but scolded him strongly for his indiscipline in not respecting normal working hours.

"No, I wasn't doing that at all," he began to excuse himself.

When the waitress realized who was speaking, she apologized immediately because she knew that the professor was not someone who wasted time on entertaining himself.

At the university, almost all the teachers and students had evacuated to the countryside. The only ones who stayed behind, under the constant threat of bombing, were the anti-aircraft squads camped on the roof and the GK1 Group working in the basement. There were many work orders. When the Communications Department asked for a machine for torpedo boats, Nguyễn Trọng Quế from GK1 agreed to take care of the request, making him late for a date with his girlfriend. It was Nguyễn Trọng Quế who, when Vũ Đình Cự was in desperate need of a polarized relay, did not hesitate to offer the group the machine with a sensitivity of 150 microamps that he had personally brought from the Soviet Union. Using this, GK1 invented a device to jam unexploded bombs that had fallen near bridges, in stores or parking lots. That allowed sappers to approach and remove the explosives from the bombs.

The long-awaited day of testing the instruments for clearing the MK-52 and MK-42 mines finally arrived. The tests took place at night at the Chùa Vẽ (Vẽ Pagoda) Port on the Cấm River outside Hải Phòng. American jets could appear at any time from the aircraft carriers of Fleet 7 stationed off the coast.

The decision to begin the tests came back to Vũ Đình Cự. In the command boat, he told the leaders of the Department of Maritime Transport that everything was ready. Slowly, one after the other, the boats advanced in order of size, the smallest first followed by the larger ones. They stopped at the points fixed by the coordinates as locations of MK-52 mines, approaching them diagonally at angles of 15, 30, or 45.

On the riverbank, Nguyễn Bính followed the signals emitted by the MK-52 mines on a screen, seeing as what distance such a boat could cause them to explode. When a mine was cleared of its explosive charge, an electric light appeared in its place. If a passing boat emitted a suitable magnetic field, the light would begin to shine.

In a second experiment, boats activated the destructive components of the mines to determine the details of the explosion. A third experiment was conducted with boats equipped with jamming devices. Subsequently, the same tests were carried out with MK-42 mines. On several consecutive nights, the results confirmed the theoretical calculations. The computer had been right.

However, these experiments had only proven successful in shallow water that was unsuitable for the larger Landing Ship Tank (L.S.T.). To destroy mines in the open ocean, a new series of tests was needed. These took place on the island of Cát Hải. When a L.S.T. approached within eighty meters of an American mine buried in the mud of a mangrove, the mine's lamp shone, signifying that it had exploded. Soon afterwards, the minesweeping ships, namely Lê Mã Lương and Quyết Thắng, tested a series of de-mining devices given the numbers GK72-2, GK72-3, and so on.

Beginning at 7:15 p.m. on December 18, 1972 and continuing through December 29, the Americans launched "Operation Linebaker 2" to systematically bomb Hà Nội and Hải Phòng using B-52 superfortresses. Each day, in addition to 140 B-52 sorties, there were 300 F-11 and 500-700 strategic bombing sorties. During these twelve days, the total tonnage of American bombs was equal to the destructive force of the two atomic bombs dropped on Hiroshima and Nagasaki.

The Hà Nội Polytechnic University was one of the targets of American strategic bombing. Since the row of buildings at number 19 had already been hit, the GK1 Group moved to work in the basement of Building A, which had been a dormitory under French colonial rule. They did not leave the school, which was the only place with sufficient equipment to carry out experiments on new defensive weapons. The team received a truckload of cabbages as reserve food provisions. Rice and salt were unavailable. Every ten and fifteen minutes, B-52 roared overhead, followed by the thunder of exploding bombs.

On December 30, a car stopped in the courtyard, near a huge bomb crater. It was Professor Tạ Quang Bửu, supervisor of the work of GK1, who ran quickly into the basement. He shook Vũ Đình Cự's hands so strongly that his glasses fell off, announcing that Nixon had been forced to stopping bombing North Việt Nam and reconvene peace negotiations in Paris. The GK1 Group had contributed its part to this victory.

The Lunar New Year of 1973 was celebrated with overwhelming joy among the population. The port of Hải Phòng was opened after nine months of blockades. Foreign ships once again took up their jobs of re-provisioning the country and serving the import-export trade. A wide-ranging campaign to supply rations and armaments to the patriotic forces in the south was prepared.

For the GK1 Group, there was no rest for the weary. While they participated in the new campaign, the problem of de-mining Vietnamese territorial waters was their highest priority. According to the Paris Agreement signed in early 1973, this task was to have been carried out by the Americans, who almost entirely refused to carry this mission out. Up through April, they had only exploded three of the tens of thousands of sea mines they had laid, despite having mobilized a special task force of 500 soldiers, 20 ships and 50 helicopters. Thus it was the Vietnamese side, who undertook clearing the remaining American mines. For their contributions to the national war of resistance, the GK1 Group led by Vũ Đình Cự was given the Hồ Chí Minh Award by the Vietnamese government.

Vũ Đình Cự belongs to the second generation of Vietnamese scientists, educated after 1945 Revolution that ended French colonialism. In 1945, he entered first grade in elementary school. During the war of resistance against the French (1946-1954), he and his family left their home province of Thái Bình to seek refuge in a place with less risk of being bombed. At the end of the war, he was admitted to the Hà Nội Teacher Training University. Graduating in 1956 with outstanding marks, he became a teacher in the mathematics and physics department of the Polytechnic University. Knowing that the knowledge he had rapidly acquired at school still had many problems and mistakes, he devoted himself to learning, dividing his time between a 10-square-meter room and the school laboratory.

Tạ Quang Bửu, the director of the Hà Nội Polytechnic University, introduced him to the field of solid-state physics and sent him to Moscow. In 1965, he defended his Ph. D. thesis in mathematics and physics at Lomonosov University. In 1967, fifteen months later, he received his doctorate of science

The brilliant defense of his thesis won him the following recommendation from the director of the institution, "It is with legitimate pride that the State University of Moscow confers the degree of Doctor of Science in Mathematics and Physics to the first Vietnamese to receive this title. We have thus fulfilled the promise made to the Vietnamese leaders who visited our school: training scientists of the highest level for Việt Nam."

Vũ Đình Cự's doctorate of science thesis, along with his other work, is well known in scientific journals around the world, particularly in the application of his theory of magnetic membranes.

NGUYỄN HỮU BẢO AND THE MINESWEEPERS OF GK2

"GK2" is an abbreviation of two related names: Ministry of Communications and Transport (G) and Hà Nội Polytechnic University (K). It refers to a group of scientists and technicians from these two organizations that collaborated during the American air escalation against North Việt Nam in order to design and build minesweepers able to detect and destroy underwater mines dropped by the American Air Force.

The Pentagon intended to paralyze Hà Nội's war effort by cutting the supply lines from the North to the South. However, America had to give up its mining operations.

Almost thirty years later, I decided to relate the little known exploits of GK2. I could think of no one better to call for this than TSF engineer Đoàn Nhân Lộ, a close colleague and friend of Nguyễn Hữu Bảo, the head and soul of GK2.

I found Đoàn Nhân Lộ in a room cluttered with televisions and electric wires of all colors in one of the numerous large buildings of the imposing Hà Nội Polytechnic University, built with the Soviet aid. To have more space and talk in quieter surroundings, we went up to the roof. From this height, we could see Thống Nhất (Reunification) Park, the biggest public garden in the capital.

He pointed out the small Phong Lan (Orchid) Peninsular where visitors come and admire orchids and all kinds or ornamental plans, red fishes, squirrels, and parrots. He said, "They tell me that during the secret negotiations with Việt Nam, Henry Kissinger came there. Assistant Secretary of State Sullivan, who accompanied him, was astonished that, in spite of the continual and intensive bombing by squadrons of B-52 flying fortresses, the gardener went on taking care of the orchids.

"He would have been much more astonished if he had known that we had tested our first model of the T5 minesweeper right there on the lake. Why was it called T5? T and 5 were the initials of the words *tháng* (month) and *Năm* (May), referring to the birthday of President Hồ Chí Minh, who was born on May 1890. The T5 was not large, no bigger than a four-person boat. It was four meters long, with no seats, and needed no pilot to guide it. Surrounded with electric wires, it was a kind of magnet. Do you see the coconut palm near the lakeside? It was just at the foot of this tree that we directed the movement of the device during the first test. Pressing different buttons, we had it advance, going back and round Hòa Bình Island, and head for the other side, near the white columns with the flowers that was assumed to be the place of the minefield. Each time that the signal lamp flickered on the minesweeper, this was supposed to show the destruction of a mine. The test succeeded and everything was in order."

He recalls his war memories and exciting young days with emotion, "At the end of the sixties, I received the mission to go to Vinh to study the means of destroying American mines on the spot. Vinh was then a major target of American air force bombing.

"Before my departure, I rode my bicycle to Ân Thi, a hundred kilometers or so from Hà Nội, where my wife had gone with her postal technology factory. She was pregnant. We agreed to name the future baby *Vinh* in remembrance of my perilous mission.

"After this visit, I joined my demining team with a calm heart. We were stationed at the river port of Bến Thủy, near Vinh. The team was composed of a few dozen of people, including hospital nurses. We lived in small thatched cottages. In the corner of a house I saw a few tombstones bearing the names of deminers who had laid down their lives during their services. Their comrades had not had time to build proper tombs. They had to see to the most urgent matters: lifting the blockade of the port, where a line of boats was waiting to sail for the river mouth or the sea. Demining by hand proved to be deadly. Yet, the atmosphere in the deminers' cottages was far from pessimistic. The girls never stopped singing.

"The chief of the team, Ninh, had a scar on his forehead caused by a demining operation. He told us, "Ever since Johnson proclaimed limited bombardment, Military Interzone 4 of which we are a part has become a pocket drawing the highest tonnage of bombs in North Việt Nam. They are dropping sophisticated MK-42 mines of the ultra-sensitive 0 and 1 types. Many of us here haven't even finished primary school. However, we learned the character of mines at the cost of our lives. At the beginning, we launched the boats at full speed near the place of the mines. We had to pay dearly for this, death and wounds were common."

"At the end of our conversation a bent-backed man appeared, leaning on his stick. He seemed extremely old, although he was only forty. He was named Vân, the deputy chief of the team. He had wounded his spine from a bomb fragment. He came to us with the following request, "Our eager prayer is that the technicians of the ministry will find a less deadly demining method. Until today, we have used wood or bamboo canoes with cooper or aluminum nails, without anything made of iron at all. Sailors strip to their waists, only wearing shorts, each equipped with a life-vest. They steer these canoes, pulling a net with a magnet behind them. The boats pass by the mines, which explode only when the net catches them. The explosion hurls the sailors up in the air, and they fall into the water. Those who are still conscious swim to the riverbank, and those who lose consciousness will float thanks to their life-jackets."

During their stay in Vinh to study demining, Đoàn Nhân Lộ and his supervisor and friend, Nguyễn Hữu Bảo, took part in these risk-filled, rudimentary operations. This strengthened their determination to design an unmanned boat to detect and destroy American mines.

The construction of a telecommanded minesweeper was a collective undertaking under the direction of Nguyễn Hữu Bảo, an electrical engineer of the Navigation Department of the Ministry of Communications and Transport. There was no problem from a theoretical standpoint. But almost insolvable questions existed in practice, beginning with the insufficiency of materials. How to set up high-fidelity transmitters and reception apparatus from an incongruous medley of accessories collected from remnants of shot-down American planes and other sources?

Experimentation took place on the southern bank of the Red River. Đoàn Nhân Lộ stayed on the roof of a subterranean blockhouse near the Museum of History (formerly called the Louis Finot Museum) with the transmitter and batteries. Nguyễn Hữu Bảo had the receiver and batteries with him, riding on the back of a friend's bicycle. The friend cycled him along while Nguyễn Hữu Bảo registered the distances he wanted to measure. The results proved satisfactory: with the material at their disposal, the telecommand could work from a distance from one to five kilometers depending on the terrain.

The construction of a minesweeper was entrusted to the Bảo-Lộ group, which included Đinh Ngọc Liễn, a mechanical navigation engineer, Phạm Văn Đương, a TSF engineer; and others. The members of the group had studied in different places in China, the Soviet Union, and Poland.

The minesweeper underwent several tests. The one that took place in the presence of Minister of Communications and Transport Phan Trọng Tuệ failed because the boat could not be telecommanded. A new test in the presence of Minister of Higher Education Tạ Quang Bửu was satisfactory.

The appearance of the T5 was delayed for a while because American President Johnson announced a break in the bombing. But this truce did not last long. In April 1972, Nixon resumed the bombing of North Việt Nam. Contrary to the hesitating manner of Johnson, Nixon wanted to deal heavy strokes, trying to destroy all land and water communication routes. He attacked Soviet and Chinese boats in the port of Hải Phòng and all the remaining bridges. He dropped 7,000 MK-52 mines to blockade twenty-five river months.

The Americans chiefly aimed at the watercourses, dropping some 6,000 MK-42 and other types of mines, preferably in the pre-dawn mist in order to make detection more difficult. The new models of MK-42 (2,3, and 4) posed new problems as their semi-automatic devices could change the direction of minesweepers.

On the nights of August 23-24, 1972, the American Air Force mined a section of the Đuống Canal, a branch of the Red River about ten kilometers from Hà Nội. The chief of the Department of River Navigation gave the order to mobilize the T5 for the first time.

The test took place on the night of August 25 in the landing strip of Bến Trì, near Chi Nhị. The demining group responsible for this river section had just suffered casualties: a TX boat equipped with a foreign-made detection device had exploded on a mine, resulting in one dead and four injured. The morale was not favorable to the experiment, all the more so as the watchers could not accurately pinpoint the places where an American plane, probably an A-3J of the American Navy, had dropped the mines during a nighttime rain shower.

The operation could not start until the morning of the next day. But at 11:00 a.m., an air raid occurred only one kilometer from the experiment spot. The T5 had to be brought back to the riverbank and camouflaged with a layer of duckweed.

At 2:00 p.m., the maneuver took place again. Nguyễn Hữu Bảo and Đoàn Nhân Lộ were anxious, fearing that if this test failed, people would no longer put trust in their T5. The commander of the operation remained on the left bank of the river, while a group went on board to reach the other bank and then walk 2 km to take their posts on the dike at a turn of the waterway. Because of the river flood, the boat could not land. The group and the material could only come ashore thanks to the devoted help of a ferryman and his daughter rowing a bamboo canoe.

The technicians waited for the T5 released on the left bank to be brought along the stream and reach a place within the reach of the telecommand. But the T5 floated towards the mined zone and had to be diverted from its direction at all costs. Fortunately, a river-dweller in his sixties offered to carry two technicians in his boat to save the T5.

All the pieces of the device had been put back in order. The T5 then exploded an American mine, raising a column of fire, water, and mud about forty meters high. Models 2 and 3 of the MK-42 had found their master.

The next day, Nguyễn Hữu Bảo and Đoàn Nhân Lộ indulged in the luxury of sleeping late. Around 2:00 p.m., the systematic demining of the region began in the presence of a large crowd of curious and overjoyed local residents, gathering along both banks. At 3:00 p.m., a mine exploded, followed by another one five minutes later and so on until the entire section of the river was mine-free and transportation could be resumed.

Through manufacturing a series of T5 minesweepers, Vietnamese engineers could cope with the intensive air mining ordered by Nixon in the last months of 1972. The simple effort of pressing the buttons was enough to telecommand the engine, something that any girl of the village militia could do easily. In October, a section of the

Red River near Vạn Điểm sugar-refinery was mined. A T5 quickly cleared the place of mines. The minesweeper was then carried by truck to Bến Thủy near Vinh. After one day's training, Ninh's team was able to demine the river port.

Nguyễn Hữu Bảo thought back to the victims of the demining group during his study period with Đoàn Nhân Lộ in Vinh. He said, "We have paid our debt to their memory!"

In the end of November 1972, the Ministry of Communications and Transport organized an exhibition of the techniques used to clear the sea mines dropped by the Americans in a 20-square-meter hall. Vietnamese GK1 and GK2 machines were placed beside American MK42 and MK52 mines. The GK1, designed by Vũ Đình Cự's team, was placed on cars or small boats in order to neutralize mines it passed by. The GK2 was the telecommanded T5 described above.

In the Paris Agreement signed in January 1973, the American side accepted a protocol to demine all the waters of North Việt Nam (sea, ports, and waterways) where they had dropped mines. To carry out this decision, meetings between both sides took places in March 1973 in the military airport of Cát Bi (Hải Phòng). The American delegation was headed by Vice Admiral McCley. American technicians made an elaborate report on the characteristics of the MK-52 and MK-42 (which the Vietnamese had already acquainted themselves with) and the means for neutralizing them. On April 9, the Americans demonstrated the use of their demining machine, which the Vietnam found too cumbersome.

The Americans assumed responsibility for sea demining, supplying the Vietnamese with devices for demining waterways. But they showed so much bad faith that finally the Vietnamese side had to do all the work, most of the time by its own methods. As a result, the efforts of GK2 received the Hồ Chí Minh Award for Science and Technology.

HÀ VĂN TẤN – A SCHOLAR OF PROFOUND KNOWLEDGE

Born in a famous Confucian family in Tiên Điền Commune, Nghi Xuân District, Hà Tĩnh Province, Central Việt Nam (also the birthplace of the great poet Nguyễn Du, who was later honored as a World's Cultural Celebrity by the UNESCO), Hà Văn Tấn soon became famous for his wide and deep knowledge of Chinese studies at the age of twenty through his proof-reading of the book *Dư địa chí* (Geographical Records) by Nguyễn Trãi (1380- 1442), another World's Cultural Celebrity.

Following that, his book *Cuộc kháng chiến chống xâm lược Nguyên-Mông thế kỷ XIII* (The Resistance War Against the Mongol-Yuan Invasions in the Thirteen Century) and his research on Buddhism in Việt Nam became widely acclaimed. However, it was his study on the archaeological evidence for the Red River civilization that earned him the Hồ Chí Minh Award.

A Prestigious Family

Even though we had been close friends for a long time, he had never told me about his family. It was not until early 2000 that I found, by chance, something about his famous family tree. In the worship hall of the Hà family in Nghi Xuân District, Hà Tĩnh Province, a red-lacquered screen with three gilded Chinese characters conferred by Emperor Kangxi, an emperor famous for his vast knowledge in the early Kang-Yong-Gan period of the Qing Dynasty in China, on Hà Tông Mục can be found. It was dedicated to him when he was chief envoy of the Lê Dynasty in Anjing (present-day Beijing) in 1703. These three characters, *nhược xung hiên*, means "modesty" and "bravery". He was flexible in diplomatic behaviors, trying to maintain good relationships between Việt Nam and China, but did not make concessions in the matters of sovereignty and territory. Emperor Kangxi was filled with admiration and esteem for his intelligent and persuasive words.

The worship hall of the Hà family and Dr. Hà Tông Mục's tomb have been known as historical and cultural sites in Hà Tĩnh Province. Hà Tông Mục was not the only person in his famous family tree who gained such a profound knowledge and had his name inscribed on the stone stele in Văn Miếu-Quốc Tử Giám (Literature Temple-National University) in Hà Nội. His scholarly predecessor was Dr. Hà Tông Trình, who held office as *Thượng thư Bộ Binh* (Minister of Defense) and then *Thượng thư Bộ Hình* (Minister of Justice) and *Tế tửu Quốc Tử Giám* (Director of the National University) concurrently during the reign of King Lê Thánh Tông. He helped this great king to write *Luật Hồng Đức* (Hồng Đức Law).

Another famous figure in the Hà family was Hà Tông Quyền, who was known as *Tham tri Bộ Lại* (equivalent to present-day Deputy Interior Minister) during the Nguyễn Dynasty and a well-known poet.

Hà Văn Đại was the member of the Hà family who took part in and passed the last Confucian examination in 1919. He got his junior doctor degree when he was nineteen years old. He was also Professor Hà Văn Tấn's uncle.

Hà Văn Tấn was born on 16 August 1937 in Tiên Điền Commune, Nghi Xuân District, Hà Tĩnh Province, which was also the hometown of the great poet Nguyễn Du.

A Young Scholar

In September 1957, after graduating from the Central School of Economics and Finance (present-day Hà Nội National Economics University), I worked for the *Thủ Đô Newspaper* (present-day *Hà Nội Mới* or New Hà Nội). Later, I was fortunately provided with a small flat in 28 Lê Thái Tổ Street, right on the shore of the Hoàn Kiếm Lake, in the center of Hà Nội. I lived with Mr. Phùng Bảo Thạch, a famous journalist before the August Revolution (1945), and Mr. Chu Thiên, a staff member of Hà Nội University.

"Where are you from?" Mr. Chu Thiên asked me one day.

"I come from Nam Đàn, Nghệ An Province," I replied.

"No wonder you look like a bookworm. Do you know Hà Văn Tấn?" asked he.

"No, I don't. During the anti-French resistance war, I studied at Huỳnh Thúc Kháng High School in Nghệ An while Tấn studied at Phan Đình Phùng High School in Hà Tĩnh," said I.

"Hà Văn Tấn is the same age as you, but his Chinese is very excellent. He is the first man who presented a report entitled 'Production Mode in Asia' in front of the teaching staff and students of the Faculty of History of Việt Nam. Then he had his report entitled 'Trở lại vấn đề totem của người Việt nguyên thủy' (Reviewing the Primitive Vietnamese Totem - translator) in the *Historical Research Magazine*. Moreover, he edited and wrote notes for the Vietnamese translation of *Dư địa chí* (Geographical Records) written by Nguyễn Trãi[1]," said he.

This knowledgeable old-aged writer's admiration for Hà Văn Tấn made me very curious about this 20-year-old man, a newly-appearing talent of social sciences.

In 1960, *Dư địa chí* was printed by the History Publishing House. I got it immediately. It was translated from Chinese into Vietnamese by the Confucian scholar Phan Huy Tiếp. Professor Đào Duy Anh must have favored Hà Văn Tấn, a 20-year-old internship assistant then, when he assigned Tấn to proofread the Confucian scholar's translation, footnote, and write the foreword. The notes took up 115 pages, four times as long as the main text (thirty-eight pages long). They were aimed at providing readers with better understanding of places that were mentioned in the main text. To complete his task, he had read a great number of records on Vietnamese geography over times, thirty old Chinese books, and sixteen old Vietnamese books, all certainly in Chinese. They included works by Chinese writers such as *Thủy kinh chú* (The Book of Rivers), *Tiền Hán thư* (The History of Anterior Han), *Hậu Hán thư* (The History of Posterior Han), and *Nam Thục chí* (Records of Southern Shu Kingdom[2]) as well as those by

1. Nguyễn Trãi (1380-1442) was a politician, a military expert, a diplomat, and a great poet. He was honored as a World's Cultural Celebrity by the UNESCO.
2. China, in the Three Kingdoms period (220-280 AD).

Vietnamese writers such as *Toàn thư* (Complete History), *Cương mục* (Summary), and *An Nam chí lược* (Concise Records of An Nam).

He also pointed out many wrong details that had been added by later generations in the existing version of *Dư địa chí.*

According to historian Phan Huy Lê, in a meeting of the Faculty of History of Việt Nam, Professor Đào Duy Anh commented on Hà Văn Tấn's proofreading of *Dư địa chí*, "Very elaborate and serious. I'm quite satisfied with the work."

In the following years, I was so busy with my work at the *Hà Nội Mới* that I couldn't keep track of his next sophisticated studies on Hòa Bình Culture, Phùng Nguyên Culture, and others. Then, his scientific article entitled "Từ một cột kinh Phật năm 973" (From a Buddhist Prayer-pillar in 973), which was published in the sixth issue of the *Nghiên cứu Lịch sử* (Historical Study), produced a stir in the public of Hà Nội. This prayer-pillar was discovered in Hoa Lư (Ninh Bình Province) and identified to have been built under the Đinh dynasty. It consisted of eight faces with Chinese letters carved on them. After years of being buried underground, the letters had faded away, so it was hard to read them. Hà Văn Tấn rewrote them, and translated them first into Sino-Vietnamese and later into Sanskrit because the original version was written in Sanskrit. Then, he analyzed the meaning of each word and each sentence. Sanskrit was well known as an old language which was not understood by many people. Only scholars of profound learning could read and understand it. Hà Văn Tấn silently learned it though a German book when he was more than twenty years old.

Through his brief article, readers could easily realize his deep understanding of Buddhism although he was only twenty-eight years old at that time.

His new book, which he had cooperated on with Phạm Thị Tâm, was also widely acclaimed. It was entitled *Cuộc kháng chiến chống xâm lược Nguyên-Mông thế kỷ XIII* published in 1968. To complete it, he had to read nearly one hundred reference books in different languages such as *Nôm* (Chinese-transcribed Vietnamese), Chinese, Vietnamese, Russian, French, English, German, Japanese, Polish, and others. Among them was the historical book entitled *Sử biên niên* (A Book of Annals) written in Persian by Fazl Allah Rasid ud-Din, a famous historian in the thirteenth century. Thanks to his research, it is presently known to us that the victory of Kiefce-Kue over Tugan, which made Tugan run for his life to Lukin-phu (Jiangxi Province, China), created a great echo in remote Western Asia right in the thirteenth century and made the famous historian of the city of Hamadhan near Teheran put it down in his Persian book of history. Medieval Persian was difficult to read and understand, especially for modern Vietnamese people.

Many rare and invaluable historical documents were collected, selected, and introduced to Vietnamese readers for the first time. President Hồ Chí Minh read the book and inquired after its two writers shortly before his death. It was a rare case in Việt Nam that this book was reprinted four times.

Vietnamese historian Phan Đại Doãn said, "During the time of evacuation in Vạn Thọ Commune, Đại Từ District, Thái Nguyên Province from 1966 to 1968, I used to talk to Trần Lê Hữu, a humorous scholar of Sino-Vietnamese. Once

I asked him, 'What can you see from our facial features?' (At that time, we were over thirty years old.) 'In terms of Chinese studies, Hà Văn Tấn can be ranked as a doctor in the feudalism period. I've spent a lot of time learning and taken Confucian exams, but I really admire his wide knowledge of Chinese studies,' replied him. These words have been spoken not only to me but also to many others."

Shedding Light on the Origin of Đông Sơn Culture

Hà Văn Tấn created excellent historical works, but they were not his main contribution. It was his 850-page book entitled *Theo dấu các văn hóa cổ* (Tracking Traces of Ancient Cultures), a collection of his archaeological articles, that earned him the Hồ Chí Minh Award.

He has dedicated most of his time and effort to discovering new things in sciences and creating great works. They might not catch much attention from the public, but they have been widely acclaimed by profound scholars inside and outside Việt Nam.

Hà Văn Tấn is one of the pioneers in the field of archaeology in Việt Nam. Before the 1960s, archaeology was not an independent scientific study in Việt Nam. It consists of two parts: excavation (on the archaeological site) and analysis (in the laboratory). The former is more important than the latter.

It can be said that Professor Hà Văn Tấn frequented all archaeological sites in Việt Nam, from the North to the South, from forests and border areas to coastal areas and remote islands. For many years, he traveled to different places on a

"Thanh Niên" bike that was made in Việt Nam with a cotton jacket and a palm hat as his luggage. His students often humorously described his luggage "rich in national identity".

He has not had any chance to study overseas; however, his scientific studies have won world-wide recognition. They have been published in France, Rotterdam, New York, Hawaii, Tokyo, and Bangkok. He was invited to be a member of the Board of Examiners of Doctoral Dissertations at Paris University 7. Nevertheless, he still has had many of his works published in domestic specialist journals to raise their prestige in the eyes of foreign scholars.

Through excavation and analysis, he and his colleagues have discovered and named some archaeological cultures in Việt Nam. The term "archaeological culture" refers to a collection of vestiges which have some features in common, are located in a certain area, and have existed for a certain duration of time. With his valuable archaeological works published in Việt Nam, he has been widely recognized by international archaeologists.

Some of his scientific articles are rather short, only a few pages long, but they still reveal his profound knowledge and provide readers with up-to-date information. Other articles are long and provide readers with overviews of archaeological issues, reviews of archaeological achievements, and new directions of archaeological research. He is fluent in different languages, including Russian, English, French, German, Chinese, *Nôm*, Japanese, Sanskrit, and Pali. He is also able to apply advanced mathematics and modern research methods into archaeology. He has never jumped at extreme statements or generalizations to conceal his lack of evidence.

For example, many scholars tried to prove that Đông Sơn Culture did not originate in Việt Nam but overseas. O. Janse tried to track its origin in the Scythes ethnic community in the vast steppe between Asia and Europe. According to B. Karlgren, this culture sprang from the Huai River Culture in China. In R. Heine-Geldern's opinion, it was created by the Tokhara ethnic community who immigrated to Việt Nam from the Black Sea.

Hà Văn Tấn did not agree upon these opinions. However, it was not easy to reject them. He needed a lot of evidence which took him tens of years to collect. Moreover, he had to apply the latest research methods to defend his arguments. Finally, in a series of scientific articles, it was proved persuasively by him that Đông Sơn Culture was the development of pre-Đông Sơn cultures, including Gò Mun, the latest, Đồng Đậu, and Phùng Nguyên, the earliest.

I am personally interested in his works on Phùng Nguyên Culture, such as *Văn hóa Phùng Nguyên và nguồn gốc dân tộc Việt* (Phùng Nguyên Culture and the Origin of Vietnamese People), *Người Phùng Nguyên và đối xứng* (Phùng Nguyên Ethnic Community and Symmetry), and *Từ gốm Phùng Nguyên đến trống đồng* (From Phùng Nguyên Ceramics to Bronze Drums). Through his articles illustrated by detailed pictures, he showed the connection between the language of ceramics in Phùng Nguyên Culture and that of bronze drums in Đông Sơn Culture.

He concluded, "No more doubts, the similarities between the art of Đông Sơn and that of Phùng Nguyên evidenced the inclusion of the traditions of Phùng Nguyên Culture in Đông Sơn Culture. Đông Sơn Culture did not originate from somewhere else but Phùng Nguyên Culture in Việt Nam."

However, he also emphasized, "To make this conclusion more persuasive, we still have many things to do. For example, we need to collect more and more documents about the ceramics of Phùng Nguyên, extend the scope of research on bronze artifacts of Đông Sơn, and find the connection between Phùng Nguyên and Đông Sơn...."

Nearly fifty years passed from when he was a 20-year-old internship assistant at the Division of History of Việt Nam headed by Professor Đào Duy Anh until he won the Hồ Chí Minh Award. During that time, he always showed his great enthusiasm for research and discovery. He has had more than 250 research works published. By 2007 when he was exactly sixty years old, a total of twenty research students, under his instruction, defended successfully their doctoral dissertations.

In *Vân Đài loại ngữ* (Small Encyclopedia of Vân Đài) the famous Vietnamese scholar Lê Quý Đôn (1726-1784) cited his favorite words from the famous Chinese scholar Wu Lai, "If a person hasn't read 3000 books and hasn't seen others' worlds, he will never become a writer". He also cited from *Thuyết Uyển*, "Scholars should read more, debate more, and write more. Of the three requirements, the second is the most difficult."

Professor Hà Văn Tấn is one of the few Vietnamese scholars who have a profound knowledge as Lê Quý Đôn expected. He has made admirable achievements partly thanks to his noble virtue. He has never been incited by trivial benefits and artificial fame.

NGUYỄN VĂN ĐẠO: GAINING SUCCESS THROUGH SELF-STUDY

Nguyễn Văn Đạo went from being a talented and virtuous student during the anti-French resistance war to a world-known mechanician with publications of 122 research works.

He successfully defended his Ph.D. thesis and then became Doctor of Science mainly through self-study. He expressed his surprise for his great achievements in scientific research during the time of evacuation and bomb attacks.

He was the first director of Việt Nam National University, Hà Nội, and was given the State Prize of Ukraine in 1996 and the Hồ Chí Minh Award in 2000. He was also elected to be an academician of four academies of sciences in the world: the Ukrainian Academy of Sciences, the Academy of Sciences of the Czech Republic, the Third World Academy of Sciences, and the European Academy of Sciences. He was a very kind-hearted man. He always helped young talents with all of his heart.

The Old School in the Palm Jungle of Hạ Hòa

Professor Nguyễn Văn Đạo was a typical example of the generation of Vietnamese intellectuals who were trained in revolutionary schools.

He was born on 10 June 1937 in Chí Tiên Commune, Hạ Hòa District, Phú Thọ Province in Northeast Việt Nam. He attended primary and secondary schools in liberated areas. When talking with the young students, he passionately recalled his old schools, "I have visited many splendid schools and have sat in many magnificent university auditoriums, but the Hùng Vương School in the jungle of Hạ Hòa always leaves me with the most ardent memories. I have visited and lived in many places all over the world, but I will never forget the time at primary and secondary schools in Phú Thọ Province."

During his time at Hùng Vương School in Phú Thọ Province, he had a great passion for mathematics. Still, he was good at all subjects.

Later, when he became a world-famous scholar, he often said to his students, "People's happiness is found not only in material life, such as cars and multi-story houses, but also in spiritual values and in a pure and lofty mind that can never be bought by money."

And then he talked about his teenage years, "During the anti-French resistance war (1945-1954), I was educated in a liberated area. The hosts there lent us not only rooms but also beds, desks, chairs, cookers, dishes, and some other things. Despite being young, my friends and I knew how to behave properly. We voluntarily cleared up the house, carried water to the homes, and tutored the younger people. It rained incessantly, making the road muddy. Walking barefoot while carrying bamboo tubes of water on our shoulders, we tried to use all our ten toes to hang on to the pulpy road to avoid falling off. On Sunday, we went into the forest to fetch firewood. We had to use a long pole with a hook to get rotten palm leaves or to cut off dried trees. Diets were poor and insufficient. Day by day, we had only rice, cassava, and cassava soup. Vegetables from the forest were boiled to dip in salt water. Due to nutritious shortages, we felt hungry all the time."

Attending Class When French Bombers "Went to Bed"

While American bombers could release star-shells and could bombard all day and night, French bombers at that time were less ultramodern. They only fought in the daytime. They targeted crowded places like three-way crossroads, markets, schools, and ferries. Therefore, students had to attend class in the evening when the bombers had "gone to bed." The class started at 6:00 p.m. and finished at 9:30 p.m. In the classroom, on two sides of the blackboard, there were two big bowls on bamboo easels containing oil which was pressed from fruits. It was lit by a wick made from the body of mollusk plants. Apart from these two lights, each student had their own light made by themselves from bamboo in the forest. They had to write on shabby and tottering tables and chairs.

They also needed another kind of light made by empty bottles with wicks inside, protected from the wind, which would light up the completely dark road in the forest when coming back home. Sometimes, they had a dried-bamboo torch.

Some subjects, which required doing experiments like physics and chemistry, they had to learn without conducting experiments. Mathematics was the least affected by such difficult situations as a good board and chalk were enough.

At 9:30 p.m., after the class on the last Saturday of month, Nguyễn Văn Đạo lit a torch, walked to the ferry, and passed the Thao River to come back home to take rice with him to the school for one month of eating. Students in small groups went during the night and got home at dawn. And then at 4:00 p.m. in the following day, each of them returned to school with a ten-kilogram rice bag on shoulders. Upon reaching the ferry, it would be pitch-dark. The French flyers would in deep sleep, so they were safe to light up the torches.

Training Himself in War and Cold Water

In early 1952, French napalm bombs were dropped down at the evacuated school's location. Fortunately, it was during summer holiday, so there were few casualties. Some houses were fired, and some local people were seriously burned.

Despite the fact that the appearance of the school in the village was the cause of this disaster, the villagers still protected the teachers and students and helped them with their needs. Almost every house in the village was open for teachers and students to live on. The hosts always offered them the best places in their houses.

Professor Nguyễn Văn Đạo said, "I stayed with Mr. Hiếu's family. He had a son named Trụ. Trụ was at the fourth grade and some years younger than me. Everyday I tutored him doing mathematics exercises that were designed by myself. He was intelligent and had good perceptions. His family regarded me as one of their members. In the evening, after studying, we used to grill cassava and boil sweet potatoes. In winter, we slept warmly in a straw bed. I wish I had been able to take photographs of the students during this time. They were difficult but happy..."

Mrs. Lê Thị Túy, a retired journalist who used to be the general manager of students at Hùng Vương School, added, "Nguyễn Văn Đạo was one of the first fourteen students to be admitted to the Young Pioneers Association and was the only student to be honored as the 'Model Student of School.' Among the members of the Young Pioneers Association those days, many became successful and held important positions in society later such as Phan Diễn, member of the Politburo of the Central Party Committee, and Chu Tuấn Nhạ, Minister of Science and Technology."

In October 1954, Hà Nội was liberated, and Nguyễn Văn Đạo went to the capital city for tertiary education.

A Great Passion for Reading and Studying

In 1957, graduating from the Faculty of Mathematics in the Hà Nội University of Education with excellent marks, Nguyễn Văn Đạo was assigned to the Hà Nội University of Technology for teaching theoretical mechanics despite the fact that his major was mathematics. The biggest challenge to him was how to understand and teach theoretical mechanics following the book *Asymptotic Methods in the Theory of Nonlinear Oscillations* written by N. N. Bogolyubov and Yu. A. Mitropolsky in Russian. His Russian and his knowledge of mathematics was not deep enough to work out its content. However, he was very fond of this book and encouraged himself to master Russian and broaden his knowledge of mathematics, especially differential equations. Thus, he chose to study by himself since no one could teach him. All the teachers in his division at that time were young (mainly at the ages of twenty to twenty-three). None of them had learned about theoretical mechanics before. They only had a few weeks to prepare for the lessons. They mainly managed to gain knowledge about this subject by themselves. Because all reference books were written in Russian and French, they had a lot of troubles with technical terms that had not been translated into Vietnamese. Therefore, the whole group spent a lot of time translating and making lesson plans. Finally, all difficulties were overcome.

Being aware of the gaps in his knowledge, he decided to pursue higher education. He attended a five-year training program at the University of Lomonosov (Moscow State University). Once he was able to read and understand the Russian language and had reached a fundamental level in mathematics, he enthusiastically started to study some new fields in theoretical mechanics. In 1960, he took the initiative in establishing with his faculty colleagues a new direction of study – the non-linear oscillation of dynamics systems.

At that time, he was staying in a cement-roofed room with soil and straw walls in the living quarter of the university. It was on the bank of the Tô Lịch River, which was muddy and closed. Mosquitoes were swarming, so he had to read books in a mosquito-net.

At the age of twenty-eight, he successfully completed and defended his Ph.D. thesis at Moscow State University.

Evacuation Area – An Ideal Place for Studying

By 1965, when the American army started to carry out bombing in North Việt Nam, Nguyễn Văn Đạo, together the Hà Nội University of Technology, evacuated to a remote village of Tày people in Tràng Định District, on the bank of the Kỳ Cùng River. There, everything was inadequate except for time and solitude, which were considered extremely valuable to those who were interested in theoretical studies.

He thought of the advice of the well-known Russian scholar M. Lavrentiev, "Are you looking for a scholar? First and foremost, find those who are able to work for such a long time, not six or eight but from ten to twelve or even fourteen hours a day."

Some people must be wondering whether it is interesting to spend a lot of time on such stressful work. As Nguyễn Văn Đạo said, science can bring a great pleasure to a person who really has a passion for it.

Teacher Trịnh Văn Quý recorded, "In the evacuation area, Nguyễn Văn Đạo used to stay up late to read and would write down what flashed into his mind. His eyes lit up whenever he was given 'priority' petroleum from Hà Nội by an army ordnance officer. As he said to me, he wished for day light so that he would not have to suffer from this indistinct light."

Nguyễn Văn Đạo used to tell me about the time he stayed in the Tày's village, "I couldn't forget the bitter cold in the highland of Lạng Sơn. However, to my surprise, it was the time when my studies resulted in great achievements. Perhaps, it was because I could focus my entire mind on theoretical study. Each day I spent fourteen hours studying and staying focused on my favorite aspects without any distraction."

In the jungle, he sometimes received letters, not from a girlfriend but from the Russian scholar V. O. Kononyenko, an academician of the Soviet Union Academy of Sciences. Such letters were sent to the address "Việt Nam, BC24ĐQ- TH2B," which stood for the Hà Nội University of Technology at that time. Here is a quotation from one of this scholar's letters to him, "I'm pleased to know that despite the fierce war, you still continue doing scientific research. Your selected topic is very interesting, both theoretically and practically."

This scholar also advised him to send their research to the *Mekhanika* (Mechanics) journal in the Soviet Union or the *Journal of Technical Physics* in Poland.

Having Mr. Kononyenko's support, he and his friends decided to send their research to some foreign journals.

It was also Mr. Kononyenko who gave his team the direction in doing their research in 1960 when visiting Việt Nam as a member of the delegation of the Soviet Union Academy of Sciences and delivering the lecture titled "Several Current Issues of Nonlinear Oscillation Theory." Then, he did his Ph.D. thesis under the guidance of this scholar.

International relationships are necessary to scientific research.

Great Interest in Doing Research in Spite of Hunger and Coldness

In the evacuation area, despite suffering from coldness, hunger, and inadequate food and clothes, Nguyễn Văn Đạo and his friends were always engrossed in scientific research. A professional refresher class for the departmental staff was held regularly. Everyday, they met to point out research issues, and then each one would go to mountain villages to deliver lectures. His lectures in this class were collected and published in a monograph book titled *Fundamental Methods in Nonlinear Oscillations*, which is still used as a reference book today.

"A little difficulty that I had to face those days was the lack of petroleum," said he. "I tried to look for a long lamp chimney that could make the light brighter and consume less petroleum."

Indeed, there was another "little difficulty" that he did not want to mention. At that time, the dinning hall of the Hà Nội University of Technology did not serve breakfast. Because he was too "lazy" to go to the mountain villages to buy bananas and cassava, he did mathematics exercises in hunger every morning.

"During the evacuation time, did you often come back to Hà Nội?" I asked.

"At that time, it was normal to cycle 200 kilometers from the evacuation area to Hà Nội for meetings. We had to set off at night. I couldn't forget the road from Đồng Đăng to Hà Nội. While going down the slope at a high speed, my eyes were half-closed. Once, I fell because I was nodding off. Fortunately, there was no injury. Arriving in Kép, which we jokingly called Kiev (the capital of Ukraine), we felt like we were reaching Hà Nội although it was eighty kilometers far from Hà Nội. After some days of meeting, we returned to Lạng Sơn. We all wished to be able to cycle leisurely during the day without bomb attacks."

Writing the Habil Doctorate Thesis in the Rain of Bombs

In 1968, America stopped bombardment in North Việt Nam. Nguyễn Văn Đạo came back to Hà Nội.

However, in 1972, America brought B-52 bombers to the north of Việt Nam for the second bomb attack, which was much fiercer than the first one. Hà Nội and Hải Phòng were covered in fire and violence. He had to evacuate to a place near Hà Nội. To make use of the time, he collected and wrote down the research results that he had recorded during the time in Lạng Sơn. These were the first pages of his Habil Doctorate thesis.

In 1976, going to Poland as a senior trainee, he took with him a complete Habil Doctorate thesis with 500 pages being deliberately studied and elaborated during the years of evacuation in the jungles in North Việt Nam. The consultant professor was extremely surprised when looking at his thesis. Three months later, Nguyễn Văn Đạo successfully defended his thesis. Coming home on the Lunar New Year's festival, his family were honored by a visit from Mr. Trường Chinh, the Chairman of the National Assembly of Việt Nam, in their modest apartment by the Tô Lịch River.

Receiving the State Prize of Ukraine

About twelve years ago, at 14.30 p.m. on 31 December 1996, Professor Nguyễn Văn Đạo received a phone call by Professor Yu. A. Mitropolsky from the capital of Keiv. He was informed that the Ukrainian President L. Kutchma had just signed the decision to award the State Prize of Ukraine to several scientific and technical works in this country. One of them was *Applied Asymptotic Methods in Nonlinear Oscillations* by Y. Mitropolski and Nguyễn Văn Đạo.

On the afternoon of a day at the end of 2006, ten days before he passed away, Professor Nguyễn Văn Đạo said to me, "It's my great honor to be awarded the State Prize of Ukraine along with well-known scholars like Yu. A. Mitropolsky. As you know, a half century ago, I was twenty years old and became a teacher at the Hà Nội University of Technology. At that time, I was reading the book *Asymptotic Methods in Nonlinear Oscillations* in Russian by N. N. Bogolyubov and Yu. A. Mitropolsky, but I could not understand it. The only way to develop the knowledge was self-study as at that time in Việt Nam there was no professor in theoretical mechanics. Surprisingly, I had been assigned to teach this subject. Theoretical mechanics was also known as pure mechanics. Yet, we called it mystical mechanics as it was so difficult and mystic."

Reading and understanding the Russian language and reaching a fundamental level in mathematics, he enthusiastically started to study some new fields in theoretical mechanics. However, at that time in Việt Nam, there was no professor to give him direction for his Ph.D. thesis. He finally decided to write to the famous Soviet mechanician whose names he saw in mechanics journal to ask for their advice on his study subjects. Then he received some replies. One scholar advised him to study celestial mechanics as some interesting issues in this field had not been solved. Having received this reply, Nguyễn Văn Đạo admitted that celestial mechanics was necessary in Russia as Gagarin and Titov had gone into space. However, he wondered if it was practical in Việt Nam since the country was still in wartime.

Ultimately, he brainstormed and selected the research subject based on his ability and the real situation of Việt Nam. Then, he received a letter from Mr. Kononyenko when he was evacuating to the bank of the Kỳ Cùng River. In it, the Russian scholar said, "Your research subject is really interesting, both scientifically and practically."

The Widespread Reputation

Nguyễn Văn Đạo had totally 122 articles, scientific reports, and professional books published. Forty-two of them were published in international journal. His name was widely known in the field of mechanics, and his works were quoted by some major mechanician in the world.

The book written in English, *Applied Asymptotic Methods in Nonlinear Oscillations*, by Y. Mitropolski and him was published by Kluwer Academic Publishers, a famous science publisher, in 1997. This 342-page book was awarded the State Prize of Ukraine.

His another book, *Stability in Dynamic Systems*, was also written in English and was published by Việt Nam National University Press, Hà Nội. This 235-page book was chosen by Professor L. Bevilacqua to be a course book at Rio de Janerio University, Brazil.

Professor Mitropolsky from the Ukrainian Academy of Sciences said, "Professor Nguyễn Văn Đạo is a renowned scientist. He has made the major contribution to the development of mechanics and applied mathematics, especially in oscillation theory.... His research results led to the foundation of a nonlinear oscillation school in Hà Nội. This was recognized in the regular conference on nonlinear oscillations held by the International Union of Theoretical and Applied Mechanics in March 1999 in Hà Nội." Professor Nguyễn Văn Đạo was the co-chairman of this significant conference.

Professor Ladislav Pust, an academician of the Academy of Sciences of the Czech Republic in Mid-Europe said that the book written by Yu. A. Mitropolsky and Nguyễn Văn

Đạo, *Applied Asymptotic Methods in Nonlinear Oscillations*, was well-known in Czech.

Professor and Doctor of Science Zbigniew Osnski from the Warsaw University of Technology in Poland said, "Professor Ngyễn Văn Đạo receives a lot respect all over the world.... He is not only typical of Vietnamese mechanician but also a true representative for Việt Nam in the world science community."

Professor Nguyễn Văn Đạo is also greatly admired by colleagues at home. Lieutenant-General, Professor, and Doctor of Science Nguyễn Hoa Thịnh, Head of the General Department of Technology of the Ministry of Defense, said, "Since 1970, I have been reading Nguyễn Văn Đạo's research works. He has made big contributions to mechanics in Việt Nam and in the world. His articles have been published in many prestigious journals both domestically and internationally."

Professor and Doctor of Science Nguyễn Văn Điệp, Vice Chairman of Việt Nam Mechanics Association asserted, "Professor Nguyễn Văn Đạo, the founder and the chairman of Việt Nam Mechanics Association, has proved his ability in assembling and uniting the forces of research, application, and education in mechanics in Việt Nam."

Professor Đặng Đình Áng, Honorary Chairman of Hồ Chí Minh .City Mathematical Society commented, "Professor Nguyễn Văn Đạo founded the nonlinear oscillation school in Việt Nam and has international stature.... He was the Editor-in-Chief of the *Việt Nam Journal of Mechanics* and a member of the editorial staff of the *Advances in Mechanics* and *Problems of Mechanical Engineering and Automation.*

He worked in a thoughtful, careful, and rational way. In fact, he is not only good at managing but also excellent in doing research. He is really outstanding both in Việt Nam and all over the world."

He always helped young talents with all of his heart. A lot of his students have been successful in their careers. They have become full professors, associate professors, doctors of science, and doctors of philosophy. They include three ministers. The most typical one is Professor Ngô Bảo Châu, who has become a professor at Paris-Sud University. Despite young age, he has won many international awards for mathematics such as the Clay Research Award, Oberwolfach Prize, Sophie Germain Prize, and Fields Medal.

As colleagues have stated, "Professor Ngyễn Văn Đạo is an outstanding talent and a kind-hearted manager."

In 2000, he was given the Hồ Chí Minh Award. He was elected as an academician of the Academy of Sciences of the Czech Republic (1988), of the Third World Academy of Sciences (1999), of the Ukrainian Academy of Sciences (2000), and of the European Academy of Sciences (2002).

NGUYỄN VĂN HIỆU – THE PIONEERING NUCLEAR PHYSICIST OF VIỆT NAM

Nguyễn Văn Hiệu was seven years old when the August Revolution broke out in 1945, ending the period of French colonialism. He thus belongs to the second generation of modern Vietnamese scientists. Years later, during a ceremony in honor of Nguyễn Văn Hiệu held at the Vietnamese Academy of Science and Technology, academician Trần Đại Nghĩa summarized his career as follows, “Born into a poor, patriotic family, Nguyễn Văn Hiệu was educated during the time of the revolution and the wars of resistance. Nurtured by the Soviet-Vietnamese friendship, he has brought honor to his country thanks to his rare talent and extraordinary energy. At eighteen, having completed his university studies, he began to teach at Hà Nội University. At twenty-two, he was admitted as a researcher at the Joint Institute for Nuclear Research at Dubna (former Soviet Union). At the age of twenty-five, he passed his Ph.D. examinations. He successfully defended his doctorate of science thesis a year later. At the age of twenty-nine, he contributed to the discovery of a new law of atomic behavior. At thirty, he was appointed a university professor in the USSR and became director of the first Institute of Physics in Việt Nam. His 120 published articles have appeared in specialized journals with global renown. He is a member of the Academy of Sciences of the USSR, of the German Democratic Republic, and of the Academy of Science of the Third World."

Trần Đại Nghĩa continued, "As an eminent researcher, Nguyễn Văn Hiệu has distinguished himself as a mentor of young scientists. He has directed the post-graduate research of dozens of students, of which many have become professors themselves. During his fifteen years at the Vietnamese Academy of Science and Technology, he has demonstrated excellent scientific organization, talent, motivation, and a spirit of responsibility. He is always devoted to his work and to the economic and cultural sectors of the country. His commitment is real, not only for form or appearance's sake. This scientist has given evidence of his deep patriotism. His success has not come about by chance. Although he attained his doctorate of science at the age of twenty-six, he has not rested on his laurels. He has continued to pursue new research, forming a group of competent co-workers who have achieved prestige in their investigations of quantum field and elementary particles. In order to link his work with the concrete situation of the country, Hiệu has also carried out more than a decade of research on solid-state theory with a group of young specialists skilled in this branch of physics."

These statements by Trần Đại Nghĩa reveal the essential qualities in Nguyễn Văn Hiệu's personality. Looking at his life in detail, this picture becomes even clearer.

The Bitterness of Youth

Nguyễn Văn Hiệu once confided to me, "During my childhood, I suffered plenty of bitterness and anxiety. Still, I never cursed my luck, since these misfortunes taught me the meaning of energy and resolution."

He was born in the village of Cầu Đơ in Hà Tây Province (presently in Hà Nội) approximately twenty kilometers from Hà Nội. His father, an artisan, joined the revolution from the beginning and became a member of a local administrative committee. When the French kindled their war of reconquest, the people of Cầu Đơ burned their own houses according to the scorched-earth policy, creating a wasteland for the enemy. Since his father went first to the mountains of Việt Bắc, then to Interzone 3 to serve the resistance, the twelve-year-old Hiệu, his mother, and seven siblings had to flee from their homeland to avoid French retribution. He had to interrupt his primary schooling three times in order to spin cotton with his mother and find some way to feed the family. Threatened by hunger, they finally rejoined his father, by then ill and out of resources.

"Fortunately," he recounted, "the leaders of the private Tống Duy Tân primary school where I was studying in Thanh Hóa Province took pity on me. They came to our home to console my parents and gave me a scholarship. Touched by their generosity, my brothers and sisters and I worked countless hours to spin cotton in order to feed our father."

He was thus able to complete primary and secondary school, moving on to the Đào Đức Thông private school that was also located near his home.

Hà Nội was liberated in 1954, after the battle of Điện Biên Phủ. He was admitted through an entrance examination to the Teachers' Training College. "For the first time in my life," he recalled, "I had time to play football and the guitar. The school housed and fed me. When it came time for the final exams, I was carried way with enthusiasm. In my eighteen-year-old imagination, I saw myself as a high school physics teacher, winning the hearts of my students through captivating lectures so that some of them could go on to be prominent physicists."

His romantic dreams of the countryside were never realized. His appointment to Hà Nội University in 1956 opened a wider door to the future. He began deepening the knowledge he had gained in secondary school and college and immersed himself in modern theories of mathematics and physics in order to reach the point of carrying out his own research in a space of four or five years.

In 1958, he was fascinated by a conference organized by Professor Tạ Quang Bửu on the discoveries of two American scholars of Chinese origin, Tsung-Dao Lee and Chen Ning Yang, who were later awarded Nobel Prizes in physics. He did not yet fully understand the significance of Lee and Yang's theories about parity violations in the weak interaction between elementary particles, but he dreamed of being able to explore this direction on his own.

Around this time, he was chosen as part of a group of young scientists to be sent to the Soviet Union for professional training. However, he declined the offer to leave immediately, asking instead to stay in Việt Nam for a while longer in order to put his belongings and scientific equipment together, as well as to find a sponsor for his thesis research. In 1960, the State Committee of Sciences sent him as a researcher to the Joint Institute for Nuclear Research at Dubna.

A Physicist's Upward Trajectory

During this period, high-energy atomic physics followed two paths of development: neutrino physics and the analytic theory of the strong interactions between elementary particles. Both of these orientations were practiced in Dubna. Before his arrival, Nguyễn Văn Hiệu had read the book *Hyperons and K-meson* by Soviet neutrino physicist, M.A. Markov. Once he was in Moscow, he requested to work in a group under Markov's direction. His first assignment was to verify the calculations and formulas in new articles by Tsung-Dao Lee and Chen Ning Yang on the K-meson particle, then present the findings in a seminar for the group. He passed this test without difficulty in passable Russian that he had learned during six years of self-study in Việt Nam. The head professor then assigned him a research subject of his own.

As he advanced in this field, he wasted no time in becoming up to date with the most theoretical research on the strong interaction between elementary particles, in particular the analytic theory of particle scattering developed by Academician N.N. Bogolyubov and Professor A.A. Logunov. From the beginning of his stay in Dubna, he never missed an opportunity to become acquainted with the most recent research. While carrying out the tasks assigned to him by Markov, he never missed the seminars given by Logunov. Once, on his own initiative, he achieved a mathematical proof of a hypothesis made by Logunov, to the great surprise of the professor. This led to his acceptance as Logunov's collaborator in spring 1962, one and a half years after arriving in the USSR. This was the turning point in his career, leading to his doctorate of science in mathematics and physics and eventually being awarded the Lenin Prize.

Working closely with Logunov in early 1963, he was able to reach his full capacity. He developed a series of theorems on asymptotic limits and published twelve articles on neutrino physics. In April, he successfully defended his Ph.D. thesis on problems in neutrino physics with Markov as his sponsor.

Immediately afterwards, he returned to study the theorems on asymptotic limits under the direction of Logunov. In July, he was invited to make a presentation on this subject at an international conference on elementary particles in Sienna, Italy. He managed this in English which he had learned by himself during his stay in Dubna. Professor Léon van Hove, Director General of the European Center for Nuclear Research, congratulated him on the research result of the Logunov-Hiệu group, findings which closely matched Van Hove's own. Van Hove invited Nguyễn Văn Hiệu on several occasions to come and work with him in Geneva and supported his admission to the International Commission on the Development of Particle Accelerators.

In Sienna, Nguyễn Văn Hiệu also met a 27-year-old Vietnamese physicist, Jean Trần Thanh Vân, who had received his State doctorate from the École Normale Supérieure in Paris. A solid friendship developed between the two young scientists. Later in life, they jointly organized a series of *Rencontres du Vietnam* international conferences on elementary particles and astrophysics held in Việt Nam in 1993, 1995, 1999, 2000, and 2006.

Back in the Soviet Union, he received permission to prolong his stay in order to complete his doctorate of science thesis. On May 4, 1964, he defended his dissertation in a brilliant fashion. Four years after his arrival in Dubna, he had completed both his PhD and doctorate of science degrees.

His doctorate of science thesis concerned asymptotic relations between scattering amplitudes in the relativistic local quantum field. According to Markov, the equations determined by Nguyễn Văn Hiệu were of great importance to elementary particles, presuming they could be confirmed by observation. (They were, indeed, confirmed forty years later.)

Russian journalist Svanev has given a detailed account of Nguyễn Văn Hiệu's doctorate of science thesis defense. "The opinion of the jury members," he wrote, "was that Nguyễn Văn Hiệu is a scholar of high caliber, full of talent, at an elevated scientific level. Academician N.N. Bogolyubov, Director of the Dubna Institute rated his doctorate of science thesis as excellent.... Markov remarked, 'Dubna created a favorable environment for Nguyễn Văn Hiệu. The rest depend on his work and his personal qualities. He is a great scholar. I think that he will be able to contribute a great deal to science.'"

Bruno Pontecorvo, a scholar of Italian origin who was also working at Dubna at that time after having renounced his American citizenship, asserted, "One could summarize the essence of this young Vietnamese scholar in two words: talent and energy. And one could add a third: productivity."

Soon afterwards, Nguyễn Văn Hiệu launched into further specific research concerning the symmetry among elementary particles. In order to achieve this, he directed a special group of researchers from various countries that had formed at Dubna, including Soviet, Hungarians, Rumanians, Poles, and Vietnamese. This research later led to the important discovery of unitarian symmetry by M. Gell-Mann and Y. Neeman.

At the same time, he collaborated with Logunov on research concerning inclusive relations among particles in high-energy physics. Logunov has just been appointed director general of the Institute of High Energy Physics in the city of Serpukhov. Nguyễn Văn Hiệu went there one week each month, covering the 225 km by train on a regular basis over a three-year period (1966-1969). The Logunov-Nguyễn Văn Hiệu research group also included M. A. Mestvirishvili, Y. D. Prokoshkin and S. P. Denisov. In October 1967, the group published the first results of its research in an article entitled "Scale Invariant Rule of Scattering Cross-Section for Particle Production." The theoretical predictions of this article would later be confirmed by an electron accelerator. It was not until fourteen years later, in 1981, that the theory of particle production was confirmed and was granted a Soviet patent.

The electron accelerator at Serpukhov was the most powerful in the world at the time. Logunov and Nguyễn Văn Hiệu participated in numerous international conferences on nuclear physics in Heidelberg, Geneva, Sienna, Vienna, Kiev, and London to share the results of their research and discuss it with their colleagues. Experimental verification in nuclear physics laboratories in these and other countries confirmed the accuracy of the Logunov-Nguyễn Văn Hiệu theory. In 1967, this theory along with a theory developed by the future Nobel Prize winners Glashow-Weinberg-Salam, was judged the most important new discovery of nuclear physics in the world.

Following the granting of the research patent in 1981, further official acclaim followed. On April 22, 1986, the Soviet Union awarded the Lenin Prize for Sciences and Technology to the Logunov-Nguyễn Văn Hiệu group for their discovery of the law of particle production. Soon afterwards, Nguyễn Văn Hiệu was admitted to the Academy of Sciences of the USSR., together with the American John Bardeen, the West German Rudolf Mossbauer, and the Belgin Ilya Prigogine.

Professor Đào Văn Đức, Director of Việt Nam's Center for Theoretical Physics (under the Vietnamese Academy of Science and Technology), underlined the importance of the discoveries of the Logunov-Nguyễn Văn Hiệu group. "This law gives great impetus to the development of high-energy nuclear physics, from a theoretical as well as a practical point of view. In particular, it has contributed to the formation of quark theory, one of the most amazing revelations of contemporary nuclear physics."

Nguyễn Văn Hiệu and Physics Research in Việt Nam

I will never forget my 1980 visit to the first modern research facilities of the Vietnamese Institute of Sciences (present-day the Vietnamese Academy of Science and Technology), directed by Nguyễn Văn Hiệu.

On the south side of West Lake in Hà Nội, the then newly-completed architectural complex lay among the rice fields of Nghĩa Đô Village, formerly well-known for the manufacture of royal edict paper. Inside a thick-walled concrete chamber, a neutron generator was installed as a gift from the Soviet Union. Thanks to group of experimental physicists, the center had been able to produce rectifier diode, voltage stabilizer diode, light-reception diode, and a mechanical-electrical convertor, which were in great demand in the market.

A number of laboratories were sufficiently equipped to allow foreign researchers to come there to work. Theoretical physicists obtained results that became known internationally in quantum fields, elementary particles, and solid body theory. Under Nguyễn Văn Hiệu's leadership, many young physicists succeeded in research in this last area, developing theories of particle diffusion, quasi-particles, fluorescence, and absorption of photons.

When he was at Dubna, he could devote himself entirely to his own research. When he came back to Việt Nam, he had to focus the largest portion of his time to the administration and management of the work of other physicists. There were innumerable annoyances to deal with: power outages, a lack of equipment, long administrative meetings, and repairs. Despite this, he never abandoned his own research. He found time in evenings, weekends, and holidays to continue working.

Appointed Director of Việt Nam's Institute of Physics upon his return from the USSR, he had to begin from scratch. He had a thatched cottage built as his office, while an old French blockhouse was converted into a place for the study of radiation chemistry. He asked for scrap iron from the Gia Lâm rail yards, which had been destroyed by American bombing, to serve as table legs.

Thanks to Professor Nguyễn Văn Hiệu's international reputation, however, his new institute did not have to wait long to receive material aid from sister physics institute in the Soviet Union, Germany Democratic Republic, Hungary, and Czechoslovakia, as well as from individual British, French and American scientists. In April 1972, when President Nixon resumed the bombing of North Việt Nam with unprecedented intensity, tons of scientific materials for the institute were stuck at the port of Hải Phòng. The institute's staff risked their lives to recover it and evacuate it to a safe location until the end of the American aerial escalation.

With human and material resources in hand, Nguyễn Văn Hiệu nurtured two ideas about the institute's development in his head. The first was to continue to follow the increasing pitch of discoveries in international physics. At the same time, he hoped to serve the concrete economic and reconstruction needs of his country. Towards this second objective, the institute collaborated with various branches and organizations: the Institute of Geology and Minerals, the Cartography Bureau, the Technical Department of the Interior Ministry, the Military Technology College, Institute of Military Medicine, and the Institute of Metallurgy. In these partnerships, the Institute of Physics conducted research on semiconductors, electronics, lasers, and other subjects.

After his return to Việt Nam in 1969, he never stopped building relationships with foreign scientists and scientific organizations. Solid relationships developed over time with the Soviet Union, United States, Italy, South Korea, and elsewhere in the Asia Pacific region. In 1996, he received the Hồ Chí Minh Award for his research on the structure, symmetry, and interaction among elementary particles and on quasi-particles. Although this research was carried out largely from 1960-1965, for him it continued to be the work of a lifetime.

NGUYỄN THỊ LÊ – A FAMOUS PARASITOLOGIST

Born in the central province of Quảng Nam in 1939, Nguyễn Thị Lê was the first Vietnamese woman to earn a Doctorate of Science in the former Soviet Union. Having made a name for herself in the parasitology of Southeast Asia, she was elected president of the Vietnamese Parasitological Society.

She was born in a revolutionary family that had long fought against French colonialism. When she was three, her father was sent to the notorious political prison at Lao Bảo (Quảng Trị Province). This was during the time when the patriotic organization of the Việt Minh, of which her father was a member, was fighting for national liberation against both the French and Japanese. He soon escaped from prison and rejoined the underground network in his home province where he led the 1945 popular uprising in the district of Thăng Bình.

The revolution of 1945, under the leadership of Hồ Chí Minh, put an end to the eighty years of French domination. For the next nine years, her parents served the cause of the national resistance against reconquest by the French. After the French defeat at Điện Biên Phủ in 1954, the peace accords signed at Geneva divided the country in two parts. The South passed from possession of the French into American control. The armed forces and cadres of the Việt Minh regrouped to the North, waiting for general elections to reunify the country – elections that never took place due to American intervention. One group of Việt Minh cadres, including her parents, remained voluntarily in the South in order to further the underground struggle for national reunification and against the Sài Gòn administration.

Thus, at the age of fifteen, the high school student Nguyễn Thị Lê left her parents to move to the North (Democratic Republic of Việt Nam) to continue her studies there. When her father visited her in 1960, during a secret mission to Hà Nội, she had just graduated from her university. After a while, her father returned to the South to carry out a difficult political and military struggle with his wife in the face of bloody repression. Reunification did not take place until 1975 with the withdrawal of the Americans and the fall of the Sài Gòn administration. Her father, who lived until he was 90, influenced her perseverance and energy, which are necessary qualities of a researcher.

Her doctoral thesis was the result of elaborate and systematic work on the parasitology of the tapeworm in the context of a tropical Southeast Asian country such as Việt Nam. This was valuable contribution to research, teaching, and medicine. She identified 305 species of tapeworms living as parasites on the birds and mammals of her country. A hundred and six of these were found for the first time in Việt Nam and ninety-one were found for the first time in their host animals. Among her specimens, she also identified one new scientific family and genus, seventeen species, and three sub-species to add to the taxonomy of tapeworms and flatworms. These results, which enriched the science of biology, needed a quarter-century of work to accomplish, from the time she was 16 until she was 41. By this time, the Soviet Academy of Sciences considered her "a famous parasitologist of Southeast Asia."

The *Dictyonograptus babeensis Nguyen* that she identified carried her family name, Nguyễn. But what about *babeensis*? This was Latinized version of *Ba Bể*, the name of a lake composed of three bodies of water. (*Ba Bể* means literally "three seas.") This immense stretch of water is located in the mountainous province of Bắc Cạn. Eight kilometers long and three kilometers wide, the lake is 20-30 meters deep and surrounded by karst mountains. This renowned tourist site is composed of numerous caves, waterfalls, steep cliffs, and islets. One year, a team of biologists of which she was a part conducted a scientific expedition in this area. They organized a shooting party. (She happened to be an excellent shot.) When evening came, the group rested in a stilt-house belonging to a Tày ethnic minority family. Her colleagues made quick identifications of the results of the hunt. She alone stayed up late at night by the light of an oil lamp, carrying dissections in order to discover the parasites that lived in the recesses, nooks, and crannies of the intestines, lungs, brain, and other animal organs.

In the liver of a wild rat (*Rattus fulvescus*), she found a strange-looking tapeworm. When it was identified in the laboratory back in Hà Nội, this specimen turned out to be part of the *Dictyonograptus* genus, of which only a single species had been previously identified in Brazil. Her discovery was thus the second species of this genus. This finding was made in 1968, but it could not be scientifically recognized until ten years later, when she was collaborating with the Soviet Academy of Sciences.

It was also Nguyễn Thị Lê who discovered the third tapeworm of this genus. This species was found inside a common sewer rat during a time of heavy American bombing in the province of Quảng Bình. She gave this species the name of *Dictyonograptus vietnamensis Nguyen*. In the mountains of Lạng Sơn Province, the site of the resounding victory over the French in the 1950 Battle of the Frontiers, she discovered a new species of parasite in the body of a pheasant. She named this flatworm after the province, *Platynotrema langsonensis Nguyen*.

She named another new species of flatworm after the first Vietnamese professor in parasitology, Đặng Văn Ngữ, who fell under American B-52 bombs during a scientific expedition to survey forms of parasitic malaria. She and her doctoral thesis advisor, Professor Sudarikov, found *Platynotrema dangy Nguyen et Sudarikov* in the body of a rhesus monkey.

Many other Latin biological names also include the words *Nguyen*, *vietnamensis*, and *vietnamense* as a result of the work of Nguyễn Thị Lê. Following in the footsteps of her predecessors who founded modern Vietnamese science, she contributed to extending the list of scientific terms linked to Việt Nam which began with the Tôn Thất Tùng Method, the Hoàng Tụy Theorem, the Tụy-type Algorithm, Tụy's Cut, the Đình Văn Huỳnh Theorem, and the eight new genii and 74 species named after Vũ Khúc. Thanks to her, the Nguyễn family name appears more than twenty times in scientific names, species, and subspecies of animals that she discovered for science. For these contributions as well as her service as professor at the Vietnamese Academy of Science and Technology and president of the Vietnamese Parasitological Society, Nguyễn Thị Lê was awarded the Kovalevskaya Prize for outstanding women scientists.

CHAPTER THREE

OVERSEAS VIETNAMESE SCIENTISTS

TRẦN THANH VÂN – A PHYSICIST HONORED BY MANY COUNTRIES

Professor Trần Thanh Vân was a native of Đồng Hới. He went to France to study at the age of seventeen. He defended his State doctorate (doctorat d'État) thesis on physics at the age of twenty-seven and became a researh director at Paris-Sud University. He established *Rencontres de Moriond* (Moriond Meetings) and was its president. He later founded and chaired the *Rencontres de Blois* (Blois Meetings) and *Rencontres du Vietnam* (Việt Nam Meetings). He has helped tens of thousands of Vietnamese students gain Vallet scholarships. He is an active member of the global physics community and has been honored by many countries.

A Hard But Ambitious Youth

Professor Trần Thanh Vân was born in 1936 in Đồng Hới, a town in Quảng Bình Province (Central Việt Nam). As a child, he was quite active. He enjoyed swimming in the sea, fishing, playing football, doing judo, photographing, playing musical instruments, and mountaineering. Through these activities, he gained abilities in different fields.

His primary school was Sainte-Marie (Đồng Hới) and his secondary school was Pellerin (Huế). In 1953, at the age of seventeen, he went abroad to France. Having only a thin short-sleeved shirt, he arrived in Lyon during a blizzard. His first unforgettable memory was of the extreme cold.

He tried hard to earn his living in Montargis, a small town, and then travelled to as far as Orléans. After receiving a baccalaureate diploma, he enrolled at the University of Sorbonne (Université de Sorbonne). He lived austerely in Paris. Being an excellent student, he was honorably awarded a scholarship to study particle physics, one of the "thorny" and key sciences of the twentieth century.

When he was twenty-two years old, he started to work at the National Center for Scientific Research (Centre National de la Recherche Scientifique) in France and prepared his State doctorate thesis at the Paris Uninersity of Education (École Normale Supérieure de Paris). At twenty-seven, he successfully defended his thesis.

The Establishment of *Rencontres de Moriond* and *Rencontres de Blois*

Since 1966, he had the initiative to cooperate with two French colleagues to hold annual meetings on particle physics called *Rencontres de Moriond* (Moriond Meetings) in Moriond Village, a small deserted village in the snow-covered Alps. These meetings attracted many theoretical and experimental physicists from all over the world, including those who had been awarded the Nobel Prize. They gathered in Moriond Village in Les Arcs, Savoie, to debate scientific issues in the morning and go skiing in the afternoon, and work again from 4 p.m. to 8 p.m. Nothing separated the "big" physicists and the "small" postgraduate. The French considered the *Rencontres de Moriond* as their national honor.

Beginning in 1989 in the Blois Royal Chateau by the Loire River bank in central part of France, Professor Trần Thanh Vân began to host a new meeting form called *Rencontres de Blois* (Blois Meetings) in order to announce and debate the newest achievements in astrophysics. Mr. Jack Lang, Mayor of Blois and Minister of Culture and Communications of France in 1989, regarded these meetings as a great pride of Blois. Thanks to it, the city was globally recognized.

The Friendship between Trần Thanh Vân and Nguyễn Văn Hiệu

Thirty-five years ago, in July 1963, at the International Conference on Particle Physics in Sienna in the north of Italy, Dr. Trần Thanh Vân (twenty-seven years old) met Dr. Nguyễn Văn Hiệu (twenty-five years old). At that time, theoretical physics was considered an "aristocratic" scientific field and only for people with "superior intelligence." Therefore, Nguyễn Văn Hiệu and Trần Thanh Vân were mistaken for talented Japanese physicists! They both felt so lonely. This must have contributed to their long lasting friendship. Their friendship resulted in the success of the six meetings of *Rencontres de Vietnam* about particle physics and astrophysics in Hà Nội and Hồ Chí Minh City. The meetings attracted physicists from more than forty countries in the world.

In 1995, Professor Trần Thanh Vân repeatedly said, "Presently our country is leading theoretical physics in general and particle physics in particular in Southeast Asia. The current problem is how to maintain that position. If we want to have a champion football team, we must invest millions of dollars in it. So why don't we invest some in our mathematics and theoretical physics teams, which are on a par with Korean teams, to keep their current positions?"

When asked about the motivation that helped him to overcome a multitude of difficulties in order to successfully hold the meetings of *Rencontres du Vietnam*, he said, "Through some international journals of physics, I noticed some young Vietnamese physicists working in many different places: Hà Nội, Huế, Hồ Chí Minh City, Moscow, Warsaw, Berlin Geneva, Paris, New York, and Chicago. I wished for these young people to have a chance to meet each other as well as well-known physicists in order to spread their pioneering ideas. From that, they could develop their own outstanding works which would foster the prestige of Việt Nam in the international physics community."

The Physics Conference in Russia to Honor Professor Trần Thành Vân

In 2001, Russia held a conference to show their respect for Professor Trần Thanh Vân. The conference was organized at the Moscow Institute of Theoretical and Experimental Physics (ITEP) from 24 to 26 May with the topic "From Smallest Distances to Largest Distances." The reports on particle physics and astrophysics were presented in the conference. These topics were hotly debated over in a number of meetings held by Professor Trần Thanh Vân throughout the years including *Rencontres du Vietnam*.

The main working place of ITEP was very palatial. It had been the fief of a duke and was given to scientists after the October Revolution.

The portrait of Professor Trần Thanh Vân smiling and wearing a blue-striped short-sleeved shirt and the English line "Tribute to J. Trần Thanh Vân" were printed on the conference icon. The letter "J" stood for "Jean" – the name he bore during nearly half a century living in France.

Professor Nguyễn Văn Hiệu said, "Professor Trần Thanh Vân has successfully held annual close meetings for the international physics community. Therefore, the French, the Russian, the American, and the Italian commemorate him and find solemn forms to honor him."

As Professor Nguyễn Văn Hiệu had said, President Jacques Chirac awarded Professor Trần Thanh Vân the National Order of the Legion of Honor (Ordre national de la Légion d'honneur), the highest order of France, in 1999. The American Physics Society recognized him as an honorary degree in 2000. The Russian Academy of Sciences awarded him the Honorary Doctorate in 2001.

Honoring the Vietnamese Physicist in Paris

We were invited to come back to Paris in the beginning of October in 2002 by Paris-Sud University.

The international conference on basic particle physics "Borders and Frontiers in Physics" took place on 4 and 5 October at the College of France (Collège de France), a globally-known center for sciences and education.

Posters featuring the French phrase "Un hommage à J. Tran Thanh Van" (To Honor J. Trần Thanh Vân) were hung in the corridor of the main hall. In 2002, at the age of sixty-six, he retired and became emeritus research director. (In France, the retirement age is sixty-five for both men and women). French and international physicists wanted to take this chance to show their gratitude to him and to acknowledge his contributions during the thirty-six years of maintaining the annual meetings of *Rencontres de Moriond, Rencontres de Blois*, and *Rencontres du Vietnam*. Those meetings, which could gather twenty-five Nobel Prize-awarded scientists, attracted international attention. Democracy and humanity were special features of the meetings held by him.

On the table of the Conference Secretariat, I received the copies of three articles which had just been printed on the morning of 4 October 2002 – the morning of the conference's opening.

The article in *Le Figaro* was titled "Showing Respect to J. Trần Thành Vân, Who Devotes Himself to Physics and Cares for Scientists."

The article in *La Croix* was titled "J. Trần Thành Vân, a humane scientist with an ability to assemble people."

Le Point wrote also an article about the famous Vietnamese French physicist.

And *Marianne* published a long reportage about his visit to Việt Nam in September 2002 written by its special envoy. It was evident that the French public cared a lot about him as he became "Character of the Day" in Paris that day.

The French Ministry of Research, the French National Center for Scientific Research, and the French Physical Society, and the French Atomic Energy Commission cooperated in organizing this conference. It attracted the participation of 223 physicists (including five who had won the Nobel Prize) from many countries and territories, such as France, America, Great Britain, Germany, Italy, Switzerland, Spain, Poland, Bulgaria, and Hong Kong (China). In this conference, they announced their new research results and showed their love for Professor Trần Thanh Vân.

The whole afternoon of 5 October 2002 was spent showing respect for him. Five Nobel Prize-awarded scientists, including Sheldon Glashow (America), Jerome Isaac Friedman (America), Jack Steinberger (Germany), Carlo Rubbia (Italy), and Georges Charpak (France), delivered speeches about his contributions to the international physics community. Professor Nguyễn Văn Hiệu, Dr. Nguyễn Mộng Giao (Việt Nam), Professor M. Danilov (Russia), and some others expressed the appreciation of the Vietnamese and Russian physics communities for him.

Along with organizational talent, he has published totally 155 scientific works in international first-class physics journals and has chiefly authored hundreds of proceedings of physical conferences. Anyone who has done research can see that these numbers are considerable.

President Nguyễn Minh Triết's Sincere Words

From 7 to 12 August 2006, the sixth *Rencontres du Vietnam* took place in the main hall at the Sofitel Plaza, a hotel by Trúc Bạch Lake in Hà Nội.

The meeting attracted 250 foreign physicists who had doctoral or higher degrees (including two Nobel Prize-awarded scientists) and 100 Vietnamese physicists.

The international and Vietnamese physics communities gathered together here to debate over new findings about two topical fields of physics: particle astrophysics and nanophysics. From this debate, priorities of research in the coming periods could be determined.

Before the opening of this meeting, in the spirit of connecting science to society, Professor Trần Thanh Vân invited the scientist James Cronin to talk to intellectuals, students, and the public in Hà Nội about matter and antimatter in the Cosmos.

Rencontres du Vietnam has been successfully held six times. These international advanced-level scientific meetings were a "golden chance" for Việt Nam to promote its image of a continually innovative, sincere, friendly, and open country.

After attending *Rencontres du Vietnam*, a number of Western physicists have had positive views of Việt Nam. *Rencontres du Vietnam* also helped many Vietnamese physicists living in foreign countries to come back to their motherland and stay connected to their colleagues. They include Professor Trịnh Xuân Thuận, Professor Phạm Xuân Yêm, Professor Đàm Thanh Sơn, Dr. Nguyễn Trọng Hiền, Professor Phạm Quang Hưng, Dr. Lưu Lệ Hằng, and many others.

President of the Socialist Republic of Việt Nam, Mr. Nguyễn Minh Triết, spent time meeting representatives of the 250 physicists from seventeen countries who went to Hà Nội to attend *Rencontres du Vietnam* in 2006. The president affirmed that the appearance of international scientists gave Vietnamese scientists new valuable findings. Moreover, this was a precious chance for Việt Nam to tighten its relationship and solidarity with other countries and enhance understandings about each other. The president also expressed his gratitude to Vietnamese scientists living in foreign countries, especially Professor Trần Thanh Vân, for their help to domestic scientists. The president expected that Professor Trần Thanh Vân would continuously recommend other Vietnamese scientists living in foreign countries to come back to their homeland to study and contribute to the development of science and technology in their own country. The president confirmed that the Vietnamese Communist Party and the Vietnamese Government always supplies favorable conditions to scientists to work and respects cooperation with other countries, organizations, and international scientists. He stated that this is the fastest route to the development of Việt Nam.

LÊ KIM NGỌC – THE FEMALE SCIENTIST WITH THE LOVE FOR ORPHANS

Born in Vĩnh Long Province (South Việt Nam), Lê Kim Ngọc has become a professor at Paris-Sud University and a world-renowned scientist in botany.

Freeing herself from the academic "ivory tower," she found and presided over a charity named Aide à l'Enfance du Vietnam – AEVN (Help for Children of Việt Nam) in France. Moreover, she made financial contributions to the construction of SOS Children's Villages in Đà Lạt and Đồng Hới as well as the Thuỷ Xuân Children Patronage Center in Huế.

Top-ranking General Võ Nguyên Giáp and his wife kept warm company with her in their house.

Being Honored in the World

According to the French press, she is first seen as a great scientist and not as a normal housewife. Her real name is Lê Kim Ngọc. She is a professor and doctor of science at Paris-Sud University and a senior researcher of the National Center for Scientific Research (Centre National de la Recherche Scientifique or CNRS) in France. However, having lived and worked abroad for a long time, she had to follow the rule "when in Rome, do as the Romans do." It meant that, like other Western women, she had to follow her husband's surname but keep her own first name. Unfortunately, her first name "Ngọc" was too difficult for Westerners to pronounce. Therefore, she employed her middle name "Kim", pronounced as "Kiêm" in her hometown. Consequently, when people came across names such as "Kiem Tran Thanh Van" (without tonal marks), "K. Tran Thanh Van," or just "Mrs. Tran" in world-top biological reviews, they hardly knew that it was Professor and Doctor of Science Lê Kim Ngọc.

The mid 1970s witnessed the continuous appearance of long articles, about 3000-5000 words, in famous magazines and journals in France, including *Le Monde* (The World), *Le Paris Match* (The Paris Match), and *Sciences et Avenir* (Sciences and Future). That these articles were all written by Mrs. Trần Thanh Vân, "a Vietnamese female scientist" and a researcher of CNRS, attracted the attention of the intellectuals in Hà Nội at that time. It was her who had discovered the laws of the blooming process and managed to control it in test-tubes. Under thrilling headlines such as "When Root Blossoms," "A Botanical Revolution," and "Forcing Orchids to Bloom in Accordance with Purchase Orders," the portrait of her with long smooth hair, a straight nose, a long flowery silk dress, and slender fingers holding an electronic microscope would cover a quarter of a newspaper page. The Việt Nam News Agency and many leading newspapers in Hà Nội at that time also published brief articles or translated versions of the aforementioned articles of this female scientist.

A Pioneering Work in Plant Biotechnology

Obviously, the foregoing popular articles only reflected the iceberg of her work and did not really give technical evaluations.

The more reliable assessment, however, came from botanical experts. She was recognized as the first person to come up with the concept "Thin Cell Layer" (TLC), marking the revolutionary development of plant biotechnology.

Not long ago, the book *Thin Cell Layer* in English was published abroad. In the introduction, Professor Gamborg from the University of Calgary (Canada) wrote, "The term "Thin Cell Layer" or TCL has become popular among us over the past three decades. It has also been associated with Professor K. Trần Thanh Vân and her brilliant research works. When it was introduced in her pioneering scientific article, TCL resulted in a revolution in plant biotechnology. The TCL method has proved its superiority over many conventional botanic regeneration and micro-propagation systems. It has also been considered to be vital in the recovery of transgenic plants."

TCL, initiated by her, has been scientifically recognized as a concept of firm values. It has been cited in hundreds of scientific publications and international specialized reviews and magazines. It is popularly employed in biotechnological laboratories in many countries, including Việt Nam.

She has written hundreds of scientific articles, the most outstanding of which are three articles published in *Nature*, one of the world's most prestigious scientific magazines.

Many of her other works have been also made public in the *Annual Review of Plant Physiology* and published by internationally-prestigious scientific publishers such as Springer-Verlag, Kluwer Academic Publishers, Plenum Press, Encyclopedia Universalis, Hachette, Gallimard, and Scientific American.

In 1973, she was invited to talk with the citizens of Paris at the Palace of Discoveries (Palais de la Découverte). Her talk was titled "How to Make Plants That Cannot Produce Flowers Blossom."

The Female President of AEVN

Despite being a globally-known scientist, she does not lock herself in the academic "ivory tower."

Feeling pity for thousands of homeless orphans, victims of long bloody wars, she and her husband wanted to help them. Yet, how could they help those children?

They realized that the importance of Christmas in the West was as great as that of the Lunar New Year's festival in the East. At that time, such technologies as telephones and email were not popular in France and America. People usually bought Christmas cards and sent them to their relatives or friends. This observation inspired this couple to select some silk pictures created by Vietnamese overseas painters, such as Lê Phổ, Lê Thị Lựu, and Mai Trung Thứ, and print them on Christmas cards for sale. These cards would not only be attractive and easily sold but also introduce the elegance of the Vietnamese. Then, they sold the cards by themselves in front of the Notre Dame Cathedral (English for Cathédrale Notre-Dame de Paris) on a chilly December night. Snowflakes were falling around the oak, birch, and plane trees that had turned leafless by the Seine River.

Many citizens in France were moved to tears at this sight and volunteered to sell cards with the couple, creating a reputable movement in France. It was then spread to America. After that, she set up AEVN and was elected as its president. This charity attracted 400 members, the majority of whom were French. She was also appointed Editor-in-Chief of the French newspaper *Lien* (Link) – the voice of the charity.

For each two-dollar package of ten greetings cards, the charity got one dollar in interest. From the accumulation of such small change, AEVN built Đà Lạt SOS Children's Village. It took one million US dollars for the construction and 100,000 US dollars annually for maintenance.

Thủy Xuân Children Patronage Center

"We do not run businesses," Professor and Doctor of Science Lê Kim Ngọc said sincerely in her warm and open-hearted accent of Vĩnh Long at the groundbreaking ceremony of Thủy Xuân Children Patronage Center on the morning of 28 August 1998. "We don't earn much. And ours is a non-governmental organization, so we cannot receive anything from the governmental budget. Therefore, we have to meticulously gather our own money and do a few small things to help unlucky children in our country. We do them with our whole-hearted love and sincerity. Money and glory can vanish like a bubble, but love and kindness remain forever."

After her speech, Mr. Helmut Kutin, President of the International Federation of SOS Children's Villages, said, "AENV is a member of our federation. I first met Mr. and Mrs. Trần in Paris when the three of us were still very young.

As famous scientists in the capital of France, they could have relaxed after work. However, these two young doctors of science volunteered to sell Christmas cards on a pavement in Paris to accumulate money for the construction of Đà Lạt SOS Children's Village and Thủy Xuân Children Patronage Center. As you can see, Mr. and Mrs. Trần and their friends with the same sense of purpose have had to sell millions of small cards! Earning money is such a hard task."

The 3500-square-meter land that had been overgrown with weeds in the hamlet of Thượng 2, Thủy Xuân Commune (in the suburbs of Huế) was cleared by a group of young volunteers with bush-hooks and sickles in the afternoon before the groundbreaking day. The ceremony was held simply but solemnly in this area. Prefessor Nguyễn Văn Hiệu, a famous Vietnamese physicist and a close friend of Professor Trần Thanh Vân, had left Hà Nội for Huế to share the happiness of this groundbreaking ceremony.

The ceremony was also witnessed by the orphans living in a low-roofed house at 108 Chi Lăng Road, kindergarteners – the orphans' second mothers, and Mr. Lê Phùng – a captain of young social workers. Soon these children would move to cozy and comfortable flats of Thủy Xuân Children Patronage Center. They would also be taken more proper care of and taught academic lessons. They would be able to play happily with and cluster around the people they would address as "mothers," "aunts," "sisters," and "brothers." Fishing pools, bakeries, and vocational classes would help them gradually learn how to earn their livings when they grew up.

In early 2001, I had a chance to revisit Thủy Xuân Children Patronage Center. Three houses had been completed. The children had moved to this new larger and nicer living place.

Professor Lê Kim Ngọc pointed at the southern hillock and told Mr. Pierre Pascal, her veteran companion in charitable activities and President of the French Federation of SOS Children's Villages, "Can you see the two-floor row of tens of flats over there? We are asking for it. We are planning to turn it into a vocational school for orphans. At first, we'll teach them how to make French-styled breads and cakes that meet the criteria of three- or four-star hotels in Huế. Next to it, a stadium, a hall, a guest house, and a house for the center director's family will be built. The director, female or male, should live in the village to take care of the children day and night. He or she is not an officer who just goes to work in the morning and leaves in the afternoon."

A Visit to the Paris-Based Head Office of AEVN

After *Rencontres de Blois* ended, I revisited Paris. One thing I did not forget to do before leaving the city was to visit the Paris-based head office of AEVN.

The head office was a small and low house located at 92 Avenue du Général Leclerc in a beautiful flower garden on a gently sloping hill. Entering it, I saw only two French female officers, a secretary and an accountant. How small the staff was!

Mr. and Mrs. Trần Thanh Vân warmly welcomed me at the office. "Seeing coquelicot blossoms in Paris, I miss peach flowers, lotuses, and chrysanthemums in our country," said Mrs. Trần Thanh Vân. "I read *Truyện Kiều* written by Nguyễn Du in high school. Although I have lived abroad for many years, I cannot forget his beautiful verses which describe the scene of spring going and summer coming: '*Đào vừa phai thắm, sen đà nảy xanh.*' (No sooner have peach flowers faded do lotuses blossom.) Then, he set pen to paper two other artistic verses, describing the rotation of seasons, from summer to autumn and from winter to spring: '*Sen tàn cúc lại nở hoa / Sầu dài ngày ngắn đông đà sang xuân.*' (As soon as lotuses fade, chrysanthemums will blossom / Long nighttime and short daytime signifies the season's change from winter to spring.) '*Sầu dài ngày ngắn*' (long nighttime and short daytime) is so meaningful."

"Ngọc is better at both French and Vietnamese literature than me," Mr. Trần Thanh Vân said, calling his wife by her childhood name. "It was with her help that AEVN managed to publish *Lien*. Just like Professor Trịnh Xuân Thuận, Ngọc studied at Western schools in Sài Gòn since her childhood, so she is more familiar with writing essays in foreign languages than me. Even in terms of scientific research, Ngọc is more successful than me. She has published hundreds of scientific articles, three of which were made known in the world top magazine, *Nature*, in England."

Lien, AEVN's newspaper in French, releases an issue for each three months and publishes articles and pictures that reflect the life in Đà Lạt SOS Children's Village, Đồng Hới SOS Children's Village, and Thủy Xuân Children Patronage Center. The newspaper also has good stories and expresses the emotions of adoptive parents in Paris after their visits to their godchildren in Việt Nam.

Mr. Trần Thanh Vân continued, "The editorial staff of *Lien* includes only one person, Ngọc. Most of the articles are written by her. The others are written by someone else but finally revised by her. I don't want to use the word 'edit.' This term sounds professional, doesn't it?"

Then they gave me several copies of the latest issue of *Lien* as a gift to bring back to Việt Nam.

Like Mr. and Mrs. Trần Thanh Vân, many other patriotic Vietnamese people who had tried to march forward into different far foreign lands have come back to their fatherland and made a lot of contributions with the common hope of together building "nine planes of hierarchy" in their beloved Việt Nam.

In August 2008, Mr. Trần Thanh Vân and Mrs. Lê Kim Ngọc were awarded the Certificate of Merit by the Vietnamese prime minister for their great contributions to Việt Nam.

TRỊNH XUÂN THUẬN – A WORLD-RENOWN ASTRONOMER

Trịnh Xuân Thuận is presently a professor at the University of Virginia in America. He is one of the few scientists who are permitted to use the Hubble Space Telescope.

His books were translated into many languages and became bestsellers in many Western countries.

Former President of the French Republic François Mitterrand invited him to accompany him to Việt Nam in 1994. Returning to Hà Nội many times to attend *Rencontres du Vietnam* (Việt Nam Meetings) on particle physics and astrophysics, Trịnh Xuân Thuận was given a warm reception by Former President of Việt Nam Trần Đức Lương.

Trịnh Xuân Thuận Speaking in Blois

Who are we? Where are we from? To answer these questions, we must understand the origin of the Earth, the Galaxy, and the Universe. Scientists in many fields such as physics, mathematics, astronomy, chemistry, ecology, biology, palaebiography, and stratigraphy in developed countries have been trying to find out the answers to those big questions. Therefore, the tenth *Rencontres du Blois* (Blois Meetings) was about the birth of galaxies, drew their attention.

I was impressed by Trịnh Xuân Thuận's talk. Blois is the town of Loir-et-Cher Province with a population of nearly 52,000 people. The topic of the meeting was "À la recherche de l'origine de l'univers: Du Big Bang à l'éveil" (Researching the Origin of the Universe: The Big Bang and After). The hall was full of audience. Many questions were asked, and everyone listened to the answers carefully. After the meeting, the speaker was surrounded by many audience members. Many French people wanted his book titled *Le chaos et l'harmonie* (Chaos and Harmony) published by Fayard Publishing House in 1998 in Paris with his signature.

A Hanoian Expatriate

Trịnh Xuân Thuận was born in 1948 in Hà Nội. He obtained his Bachelor of Physics at the California Institute of Technology (Caltech) in June 1970 and then his Ph.D. degree in Astrophysics at Princeton University in May 1974. He has been professor of Astronomy at the University of Virginia since 1990. He is also a frequent guest lecturer in many countries such as France, England, and Denmark. He was given the American Astronomical Association's Henri Chrétien Award in 1992. He has published more than 270 research works. Particularly, his eight thick books printed in France and America made him famous in sixteen countries.

His two books, *La mélodie secrète* (The Secret Melody) published in 1988 and *Le chaos et l'harmonie* (Chaos and Harmony) published in 1998, became bestsellers in France. In Paris, many newspapers wrote about these books, especially the latter. In 478 pages he tried to summarize "everything that humans know" from ancient times up to now in different fields such as physics, mathematics, and biology in order to reach new philosophical concepts. A discussion between him and Hubert Reeves, a famous French astronomer, was presented in a feature in *L'Express* on 5 February 1998.

After knowing his life of science research, I understand why so many Frenchmen in Blois waited in line to get his signature.

His books were famous not only for their scientific accuracy, but also for their poetic and philosophical language. They were written in beautiful and precise French.

Giving a Series of Lectures in Hà Nội

In the morning of 13 July 2000, a summer class in astrophysics and cosmology was held for gifted students in natural sciences at Việt Nam National University, Hà Nội, within the framework of *Rencontres du Vietnam*. Summer and winter classes in particle physics, astrophysics, and biology have been opened regularly for the past few years to encourage young students and scientists in Việt Nam.

Within ten days, Vietnamese students had opportunity to meet and talked to two world-famous astronomers, Professor Trịnh Xuân Thuận from the University of Virginia (America) and senior researcher Ludwik Celnikier from Paris-Meudon Observatory (France). They also attended their lectures, which were both condensed and suggestive.

In 1994, Former President of the French Republic François Mitterrand paid a visit to Việt Nam. He invited Trịnh Xuân Thuận and Trần Văn Khê to come with him. He considered these professors as a "friendship bridge" between the two countries. Trịnh Xuân Thuận said, "At that time, because of accompanying the President, I couldn't meet my relatives in Mai Lâm Village, Gia Lâm District, let alone students in Hà Nội. This trip is different. It is very comfortable."

He is a native of Hà Nội. However, because of historical circumstances, he only lived in Hà Nội for six years.

Before starting his lecture, he said, "I'm sorry. I studied at a French school in Sài Gòn (present-day Hồ Chí Minh City) and then went abroad for higher education, so I am not familiar with physical terms in Vietnamese. Today, please allow me to lecture in English."

Ten minutes later, he stopped and asked, "Do you understand?"

"Yes," the students answered.

"Any questions?"

Many students raised their hands timidly. Although they had difficulty in expressing their views in English, they continued to discuss and ask about astronomical issues such as the mysterious black holes, the endless number of stars in the Galaxy, the galaxies' transfer to the red side, the invisible dark matter, and the appearance and disappearance of stars. Trịnh Xuân Thuận stated that he believed in Việt Nam's youth.

An Unexpected Meeting

A few years later, in early August 2004, I met Trịnh Xuân Thuận by chance at a grilled fish restaurant in the middle of Hà Nội's ancient streets. "I saw the list of the international advisory board of the coming fifth meeting of *Rencontres du*

Vietnam, and I saw your name. Will you come back to attend that meeting?" I asked him.

"Yes," he replied. "It is one of my main tasks during nearly one month in Hà Nội. Besides, I have to do something important. I will discuss with some Vietnamese leaders in charge of science and education about the cooperation among universities in America and Vietnam. I will have a talk with students in Hà Nội about the topic "Tracking the Origin of the Universe: The Big Bang and After" at C2 hall at the Hà Nội University of Technology. I had a similar talk with students in Hồ Chí Minh City. This time, I come back due to Professor Nguyễn Xuân Hiệu's invitation at the Việt Nam Academy of Science and Technology. Therefore, I have a lot of free time to stroll along Hanoi's ancient streets and enjoy grilled fish and Thang noodles. Grilled fish with some fennel and belostomatid essence is so great!" He laughed. "Today, I have just visited my village, burnt some incense sticks at the ancestral alter, and met some relatives and neighbors. I'm very happy. My village is located in Mai Lâm, a suburb of Hà Nội, seen from the Đuống River."

I was surprised by his voice because he still spoke Vietnamese perfectly with a native accent of Hà Nội. After the Geneva Treaty was signed in 1954, Việt Nam was divided into two parts at the seventeenth parallel. Therefore, he had to leave Hà Nội for Đà Lạt and then Sài Gòn. He was only six at that time. In Sài Gòn, he studied at a French school named Jean-Jacques Rousseau (also known as Chasseloup-Laubat before the August Revolution). During his student life in Việt Nam, he had to learn all the subjects in French. Thanks to it, he could write up thick books, which are world-famous for its scientific accuracy and poetic language, later.

"I obtained my Bachelor of Physics at the California Institute of Technology, defended my Ph.D. thesis at Princeton University, and then taught for many years at the University of Virginia. It means that I speak English from the dawn to night, but I cannot speak English as fluently as French – the language I learned when I was small. My English is just enough for me to complete precise professional research with limited scientific terms. It is not enough for me to create or express something literary and philosophic," said he.

Internationally Appreciated Works

In 1992, Trịnh Xuân Thuận's book titled *Le destin de l'Univers: du Big Bang à l'éveil* (The Destiny of the Universe: The Big Bang and After) was originally written in French and published by Gallimard Publishers in Paris. In 1993, this book was translated into English by Harry N. Abrams, was released in America, and quickly became a bestseller in New York. That year, Oxford University Press also released another version translated by Storm Dunlop which was released first in England and later in America, Canada, Australia, New Zealand, India, and other countries. Thus, his book was translated into both American English and British English. Although he worked at the University of Virginia for many years, he did not translate his own book into English.

In 1993, this book was translated into German by Ravensburger and published in Munich and in other cities in Germany. Then, it was released in China (1993), Sweden (1994), Italy (1994), Japan (1995), Korea (1995), and other places.

Much later, Việt Nam began research in this field. His book *La mélodie secrete* (The Secret Melody) was translated into Vietnamese by Phạm Văn Thiều and published by Science and Technology Publishing House, Hà Nội, in early August 2000. This Vietnamese version was considered to be a gift in welcoming the opening of *Rencontres du Vietnam* with his preface in Vietnamese, "Despite historical changes, Việt Nam always prizes its education and knowledge above all the aspects of cultural life. I hope that this book will partly contribute to the spread of knowledge. I am very grateful to Professor Trần Thanh Vân for his effort which has helped this book to be published in Vietnamese. I also give thanks to translator Phạm Văn Thiều for his excellent translation. I expect this book, which was appreciated in western countries, to nurture thoughts and change the way that people think about the world. I hope that it will create scientific directions for students and that the seeds sown in this book will grow to become fruitful trees one day."

Recalling his childhood in Việt Nam, he said to Vietnamese readers, "During my childhood and adolescence, in the late 1950s and early 1960s, when I lived in Sài Gòn, one of my hobbies was being sunk in a good science book. At that time, I forget everything around me and let the author take me from my small world to a great one. Like Sherlock Holmes's investigation, I was excited about the process of scientific discovery: from signals to hypotheses, wrong directions, blind alleys, and discussions to find the truth."

He talked about his penchant that grew at young age, "The books that I read during my childhood nurtured my imagination and formed my thoughts. Surely they played an important part in leading me to science. They encouraged me in taking an active part of the great journey of science.

Then I continued to observe the Universe by telescope on the ground as well as in space."

Trịnh Xuân Thuận is one of the few astronomers who are permitted to use the Hubble Space Telescope.

When talking about his book *La mélodie secrete*, he said, "This book is about the history of the Universe. It describes in details the formation of things in the Universe, from the smallest to the greatest. It also discusses the extremely-accurate adjustment of the Universe for the appearance of thoughts that can grasp its gorgeous beauty and organization as well as assign some value to it. It also establishes the connection between humans and the Universe. We are all created from dusts like stars, so we and stars are brothers. Similarly, wild animals and wild flowers are our relatives."

This book quickly was purchased by Vietnamese students. It was reprinted in the following year, 2001. In early 2003, his book *Le chaos et l'harmonie* was translated into Vietnamese by Phạm Văn Thiều and Nguyễn Thanh Dương and published by Science and Technology Publishing House.

Returning to Hà Nội in August 2004, he met Phạm Văn Thiều, who was a physicist but who was also excellent at literature, to skim through the Vietnamese translation of a book written by him and Matthieu Ricard. This book, *L'Infini dans la paume de la main* (The Infinity in the Palm of the Hand), was published in Paris in 2000 and became a bestseller in France. It explored Buddhist teachings and the latest issues of modern science as well as shed light on secrets in the microcosm, macrocosm, and human society.

"Will you present a new research at the fifth *Rencontres du Vietnam*?" I asked.

"Yes, I will report the results we have gained in researching the formation of galaxies thanks to observations made with the Hubble Space Telescope," he answered.

I have not seen him since the fifth *Rencontres du Vietnam*. However, through his website, I am still aware of his new research projects. Every time I have difficulty in explaining modern astrophysical achievements, I turn to his books, which were written in a simple, clear, vivid, and poetic language. They are my companions in life.

In 2009, Professor Trịnh Xuân Thuận was awarded the UNESCO's Kalinga Prize on popular science and the Award of the French Academy of Sciences for his book *Les chemins de la lumière* (The Paths of Light) for its philosophical value.

NGUYỄN TRỌNG HIỀN – A SPACE INVESTIGATOR IN ANTARCTICA

In the middle of July 1994, the comet Shoemaker Levy 9 collided into Jupiter. Columns of fire burned at heights of thousands of kilometers. Pieces of the comet as huge as mountains crashed into Jupiter's surface, creating holes as large as the Earth.

Only the American Hubble Space Telescope could thoroughly observe this event which produced a stir in the Solar System. However, one Vietnamese observed this event from the beginning to the end. This person was Nguyễn Trọng Hiền. He is a scientific leader at the Amundsen-Scott observatory station in the Antarctic where twenty-seven American scientists and technical experts are working. He is also a member of an infrared physics group at the University of Chicago and a member of the Center for Astrophysical Research in Antarctica. Thus, in two programs in the afternoon of 21 and 28 July 1994, BBC London broadcasted an interview with Nguyễn Trọng Hiền on the telephone. Audience listened to his voice echoing from the Antarctic.

Success in America but Still Concerns about Vietnam

Nguyễn Trọng Hiền was born in 1963 in Đà Nẵng, near the Hàn River. After graduating from the high school, he went to America when he was eighteen years old. Overcoming the terrible difficulty of the language barrier and being good in mathematics, he was confident enough to take the entrance exam into the Faculty of Physics in Berkeley University – a prestigious university in the city of San Francisco. After obtaining the Bachelor degree, he unhesitatingly enrolled for a Ph.D. degree at Princeton University where Albert Einstein used to teach and conduct research. His major was cosmic background radiation. A few years after receiving his doctorate degree, he completed some projects in astronomy and astrophysics at the University of Chicago. Recently, he has become a researcher at the National Aeronautics and Space Administration (NASA). He also a lecturer and has been invited to teach physics at the California Institute of Technology (Caltech).

In spite of working in America, he has always had a strong attachment to his country. He has returned to Hà Nội and Hồ Chí Minh City several times to attend the *Rencontres du Vietnam* (Việt Nam Meetings) on particle physics and astrophysics.

He enthusiastically collaborated with Vietnamese astronomers to establish a modern telescope in Việt Nam. According to him, if an 85-million-people country does not have modern telescopes, youngsters are at an extreme disadvantage.

Not long after the Shoemaker Levy occurrence, on the last ten days of October 1995, the second *Rencontres du*

Vietnam was held at Thống Nhất (Unification) Palace in Hồ Chí Minh City. On 24 October, the total solar eclipse could be seen in Sông Bé (in Southeast Việt Nam). According to the assignment of Professor Trần Thanh Vân, Nguyễn Trọng Hiền along with his American workmate compiled the book "Ngày hai đêm" (Two-Night Day). This was published in a timely fashion and brought new knowledge to Vietnamese readers about this amazing phenomenon.

Nguyễn Trọng Hiền's latest research work is on the radiometer detector system, a device at the Herschel Space Observatory. This space station is named after the German English astronomer William Herschel (1738-1822). It is in a co-operative program between NASA and ESA (European Space Agency) that was sent into orbit in May 2009.

Our First Meeting in Việt Nam

I first met Nguyễn Trọng Hiền at the end of 1993 when he returned to Hà Nội to attend the first *Rencontres du Vietnam* on particle physics and astrophysics. It was held by Professor Trần Thanh Vân, Professor Nguyễn Văn Hiệu, and some Vietnamese physicists. At that time, America hadn't removed the embargo on Việt Nam yet. Overseas Vietnamese people like Nguyễn Trọng Hiền who wanted to go home had to do entry procedures at Vietnamese embassies in the other countries such as France or Thailand. It was extremely inconvenient and time-consuming. However, Nguyễn Trọng Hiền dealt with the difficulties and he was in Hà Nội on the opening day of the meeting. That year, he was thirty years old and was about to present his Ph.D. dissertation. At *Rencontres du Vietnam,* he reported his observations on cosmic background radiation in the South Pole.

After that, he and I often kept in contact with each other through e-mails. In 2003, I went to America. He helped me by introducing me to Professor Trần Duy Trác, one of his close friends at Johns Hopkins University in Baltimore.

We travelled a several-hundred-mile road to the harbour city of Baltimore and to Johns Hopkins University where Professor Trần Duy Trác was working. I was very happy when I found out that he was only thirty-three years old, but he was already an associate professor in a prestigious university. According to Nguyễn Trọng Hiền, he and Professor Trần Duy Trác have a strong attachment to each other. Being together causes them to feel less lonely in the strange land. They always encourage each other and try their best to contribute to the development of their fatherland.

Observing the Universe for Six Months

After successfully defending his Ph.D. dissertation in the honourable Princeton University, Nguyễn Trọng Hiền returned to the Antarctic for a second time in 1994 and stayed here for one year.

Before him, another Vietnamese person had gone to Antarctica. Nevertheless, Nguyễn Trọng Hiền was the first Vietnamese to explore this "ice continent" as a researcher. During his time here as the scientific leader of the American Amundsen-Scott observatory, he literally walked to the starting point of the Earth's datum line. He came with a group of Americans. They travelled on a C-130 airplane equipped to land on the ice. This was a military aeroplane which Americans had used in their war against Việt Nam. Due to the low temperature's influence on mechanical parts and machine oil and the isolation of the Antarctic, civil airlines have not taken the risk of arranging flights to this region. The only means of transport is this military aeroplane.

Being an astrophysicist, he was arranged to sit in the control room. He vacantly watched icebergs bobbing and drifting in Ross Bay. The town of McMurdo gradually appeared. In the summer, the number of residents in this town reaches to the thousands. McMurdo is the most convenient place in this continent. Here there are some modern laboratories, and it is possible to make phone calls to other continents. In the summer, ice-breakers make way for large ships carrying heavy devices and materials which are difficult to carry by planes. However, in the winter, the population reduces. A few hundred people are left. He was among that very little number of people.

After the eight-hour flight from Christchurch, New Zealand, he arrived at McMurdo, the most important seaport of the continent, located in the south of Ross Bay. He stayed here one day and went around and watched the penguins wandering on the icefall to feel less bored. The next morning, he took another flight to the Amundsen-Scott observatory.

Weather in Amundsen-Scott is harsher than in McMurdo. Plants and animals cannot survive in the extreme conditions. Yet, this is the ideal place for astronomic observation. For half a year, the sun cannot be seen which allows people to observe the Universe twenty-four hours a day.

Due to low temperatures, great heights (the observatory was located on a mountain covered with 3000 meters of thick ice), and low air pressure, the radiation here, from near-infrared waves (two microns) to hyper-frequency waves (two millimeters), are both at minimum levels. They are 200 times lower than the levels in observatories in Hawaii or Chile.

With hyper-frequency telescopes, he looked for "brown dwarfs." These are stars that are becoming smaller. They are not heavy, solid, or hot enough to start a nuclear reaction, create thermal radiation, or light. The existence of "brown dwarfs" is inferred from the stars' law of evolvement and is a vital part in discovering the solution for dark matter which is in the neighbouring region of the Milky Way. Dark matter or invisible matter occupies about 90-98% of the total weight of the Universe. This type of material does not send out light or any radiation. Consequently, we cannot see it. Yet, its appearance is obviously demonstrated by its gravitational force's impact on stars and galaxies.

The project of Nguyễn Trọng Hiền and his collaborators also involves carrying out observations on the "elderly" galaxies. According to modern cosmologists, these galaxies were shaped relatively early on in the history of the Universe. Measuring their properties of light and distribution in the Universe will provide materials to build a model of galaxies' evolvement. With advances in infrared technology, Nguyễn Trọng Hiền believes that he will achieve more satisfactory results than the previous efforts of other pioneering scientists.

Letters from Antarctica

In a letter from the Amudsen-Scott observatory in McMurdo, Antarctica, Nguyễn Trọng Hiền told me about his routines,

"Every morning, I get up very early. It is called the morning; however, it is completely dark because six successive months in the winter are nights! I enter the canteen, toast bread, drink a cup of cold water, then put on gloves, a hat, warm clothes, and go to the observatory.

Every morning is the same! The road in the pitch-dark night is nearly one kilometer long. Sometimes, I use the flashlight. Sometimes, I head directly for the light of the observatory on the rough snow surface. The snowshoes wrapped tightly around my feet are very uncomfortable. On windy days, blowing snow darkens the sky, and the view is not beyond five meters. I follow the cable that links Winter Arch – where we live – and the observatory. Wind and snow slap my face violently.

Many days, the temperature is often at -73°C. On windy days, it can reach -100°C. I've gradually gotten used to this. Sometimes, during the long route, I don't think of anything at all. Yet, sometimes, I stop and look at the surroundings and adore the mysteriously breathtaking beauty of nature....

...The South Pole is fearfully quiet. Sometimes, when feeling overwhelmed with sorrow, I relax by thinking of the telescopes which my friends and I in our country promised to design."

In September 1994, he sewed a four-meter-square red-yellow pentagram flag by himself and stuck it beside the flags of America, Russia, England, Germany, French, Japan, Australia, New Zealand, Chile, Argentina, and South Africa. This was the first time that the Vietnamese flag has fluttered at the Directorate of Protocol in Antarctica.

Hardships and Prolonged Separation

Nguyễn Trọng Hiền arrived in Antarctica in the summer and was still content on staying there during the winter despite the fact that sunshine is not seen for six months.

"A few weeks before coming to the frost region, I returned to Việt Nam after years of being far away from home. I had gone to Antarctica, but I still thought of moments with my relatives, the moments I silently watched the storks in the green paddies, or the dripping dark sky in the city I extremely loved," said he.

He always takes opportunities to return to his country. He has attended the *Rencontres du Vietnam* six times. The last time was in August 2006 in Hà Nội. As usual, after working days, he went to his hometown on the bank of the Hàn River and visited his mother, relatives, and friends. He still hopes that he will return to settle down in his hometown someday.

Former Vietnamese President Trần Đức Lương with typical Vietnamese physicists (from left to right): Professor Trần Minh Tâm, Professor Trịnh Xuân Thuận, Professor Đàm Thanh Sơn, Professor Trần Thanh Vân, Professor Phạm Quang Hưng, and Professor Nguyễn Văn Hiệu.

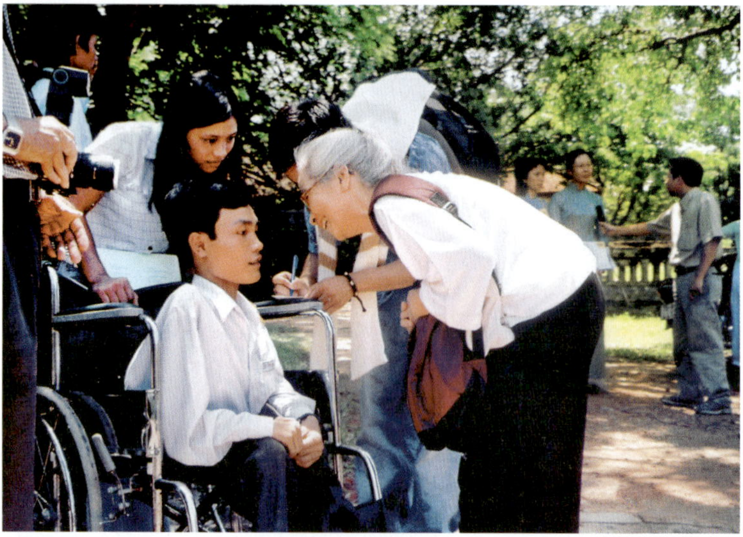

Professor Lê Kim Ngọc with a student who is disabled but good at study in Huế.

Dr. Nguyễn Trọng Hiền next to a row of doctoral steles in the Hà Nội Temple of Literature.

President Nguyễn Minh Triết receives Professor Trần Thanh Vân
in the Presidential Palace (2006).

Professor Trịnh Xuân Thuận (middle), Professor Phạm Quang Hưng (right), and journalist Hàm Châu (left) in *Rencontres de Blois* in 2010.

Professor Ngô Bảo Châu receives the Fields Medal from Indian President Pratibha Devising Pratil, August 19, 2010. He is the first mathematician from developing countries who has won that top scientific honor.

ĐÀM THANH SƠN: FROM GOLD MEDAL IN THE INTERNATIONAL MATHEMATICAL OLYMPIAD TO NEW DISCOVERIES IN PHYSICS

Đàm Thanh Sơn, who won the gold medal in the International Mathematical Olympiad (IMO) in Prague, Czech Republic, with the absolute score of 42/42 in 1984 at the age of fifteen, is now a professor at the University of Washington in Seattle.

In early 2005, he, along with two colleagues, published an elaborate work about the ten-dimensional black hole in the *Physical Review Letter*, the leading international physical magazine. This has gained a great reputation in the American and international physical community.

According to some popular influential magazines such as the *New Scientist* (April 2005) and *Physics Today* (May 2005), this scientific discovery may create a universal law in physics if it is proven completely by experiments. The monthly magazine for the International Physics Community, *Physics Today,* asked Đàm Thanh Sơn, a leading physicist, to publish a feature article to introduce his group's work.

A Great Reputation in the Field of Physics

In early 2005, Đàm Thanh Sơn from the University of Washington (US), and his colleagues, P. K. Kovtun from the University of California (US), and A. O. Starinets from the Perimeter Institute for Theoretical Physics (Canada), published a new work called *Liquid Black Hole in Ten-dimensional Space* in the world's leading physics magazine, *Physical Review Letters* (Volume 91, pp. 111-601).

This discovery almost immediately became a widespread phenomenon in scholarly circles. Popular influential scientific magazines like *New Scientist* (April 2005) and *Physics Today* (May 2005) have printed articles about this scientific work which may have spawned a new universal law in physics if it is completely proven by experiments.

The monthly magazine for the International Physics Community, *Physics Today*, asked Đàm Thanh Sơn to publish a feature article to interpret this new issue. It is entitled *Liquid Universe Hints at Strings* and is very popular on the internet.

An article by Jenny Hogan named *Exotic Black Holes Spawn New Universal Law* is printed in *New Scientist*. The word "exotic" is used by the author because the holes here are not certainly astrophysical black holes. Instead, they are just "black holes" which are modeled through string theory in ten- dimensional space by Đàm Thanh Sơn's team to describe a strong interaction fluid – quark-gluon fluid. This fluid exists in the familiar three-dimensional space. Hence, the "black holes" in the discovery are just mathematical tools which are relevant to calculations.

Jenny Hogan wrote, "That black holes may define the perfect fluid suggests a study of black holes that only exist in a theoretical ten-dimensional space. The finding may have spawned a new universal law in physics, which puts constraints on the way fluids behave in the real world.

Đàm Thanh Sơn from the University of Washington, USA, and his colleagues used string theory to model a ten-dimensional black hole as a liquid. String theory tries to explain fundamental properties of the Universe by predicting that seven dimensions exist on top of the known three spatial dimensions. While the concept is currently unproven as a cosmological model, the tools of string theory can sometimes provide answers to real problems.

It means that while the 'black hole' modeled by Đàm Thanh Sơn are not astrophysical black hole's surface areas: viscosity, which describes how thick the liquid is, and entropy density, which is a measure of the internal disorder, Đàm Thanh Sơn's team found that the ratio of these two properties is a constant which can be expressed as a mixture of fundamental constants from quantum world..."

In November 2005, in the *Scientific American* magazine, Juan Maldacena, a famous Argentine American physicist, gave an important scientific overview. In it, after discussing the discovery of black holes by Stephen W. Hawing from the University of Cambridge (UK), he mentioned Đàm Thanh Sơn's discovery of the liquid of a "newly-born universe."

He wrote, "Đàm Thanh Sơn from the University of Washington and his collaborators studied a quantity called the shear viscosity, which is small for a fluid that flows very easily and large for a substance more like molasses. The found that black holes have an extremely low shear viscosity – smaller than any known liquid."

More recent is the article entitled "Big Bang Machine" by Tim Folger in *Discover* that I mentioned in the first part of this article. To my surprise, it was Tim Folger who interviewed Đàm Thanh Sơn and gave an account of this initial discovery along with many other famous physicists' comments about this discovery.

He wrote,

"One property of quarks and gluons that Đàm Thanh Sơn and his colleagues wanted to calculate was viscosity. Using a black-hole model, they predicted that quarks and gluons should have almost zero viscosity. When experimentalists at Relativistic Heavy Ion Collider (RHIC) finally crunched through all their data, they confirmed that the quark-gluon fluid indeed had a low viscosity, at or near the theoretical minimum value predicted by the five-dimensional black-hole model…

So does this success bolster the idea that string theory is the right way to unify all of physics?

'Absolutely,' says Horatiu Nastase of Brown, who has also sought to understand RHIC's results in terms of black hole. 'At least, that's my interpretation and the interpretation of other people. My understanding is that one is experimentally testing, in this indirect way, string theory.'"

The fact that Vietnamese informational and scientific magazines such as *Physics Today* and *Scientific Activities* broke news about Đàm Thanh Sơn' s discovery in a timely and respectful manner seems to be a positive sign.

Đàm Thanh Sơn's Theoretical Prediction in Accordance with the Experimental Result

In the last five years, researchers at RHIC of Brookhaven National Laboratory in Upton, New York, have created a material at an extremely high temperature. This material had not previously existed on Earth. The experiment aims at re-generating the state that existed in the first ten microseconds after the Big Bang, the occurrence that gradually formed our universe. In those microseconds, quarks and gluons still existed in a plasma state. Before the universe cooled down, these materials had not yet combined together to become protons, neutrons, atoms, molecules, or life forms. According to the familiar quantum field theory-based calculation, the material created must be like gas, but in fact it is actually liquid.

Using string theory in a ten-dimensional space, Đàm Thanh Son's team managed to calculate accurately that the material created by RHIC is completely liquid and is nearly perfect. Its viscosity ratio and entropy density are constant. It is also relevant to basic constants in the quantum world such as Planck and Boltzmann constants.

The success of Đàm Thanh Sơn's team is that they know how to use string theory to interpret properly a difficult issue in high energy experimental physics. Brookhaven's results, stated at the meeting of the American Physical Society in Pampa, Florida, in April 2005, paid attention to the compatible calculation of string theory. This was the first time string theory was introduced in a big experiment's report. Hence, it is said that string theory is an effective tool that can reunite four interactions in nature. In other words, it could reunite the entirety of physics.

What Đàm Thanh Sơn has achieved in America shows that Vietnamese people can not only perceive the most sophisticated and complicated knowledge of modern sciences, but also can reach basic discoveries if they have the chance to work in advanced scientific environments.

A Prodigy of Hà Nội

First and foremost, we should learn about Đàm Thanh Sơn's passion for sciences that developed when he was very young.

He was born in an intellectual family. His father, Đàm Trung Bảo, is a pharmaceutical professor and his uncle, Đàm Trung Đồn, is a physics professor.

In the summer of 1984, Đàm Thanh Sơn won the gold medal at IMO with an absolute score. That year, as a scientific journalist, I arranged to meet Mr. Đinh Văn Thành, the former principal of Bà Triệu primary and secondary school to learn about Đàm Thanh Sơn. Mr. Đinh Văn Thành said,

"When Sơn was at Bà Triệu primary and secondary school, I was the principal. In the morning of the first day of school, I was sitting in my office when Ms. Lan Anh, the form teacher of class 2C (equivalent with grade 3 at present), came into my office and said, 'Sir, there is a student who has a great passion and special aptitude for mathematics in my class.'

'Lan Anh, you should pay special attention to him and foster him. When the district's competition for gifted students is held, he will be a representative for our school,' I told her.

'But sir,' she said, 'this is an especially talented student. It is unbelievable that he can apply algebraic methods in solving my arithmetic exercises.'

'Oh, really? Have you tried measuring his level of mathematics yet?'

'Yes, sir. His level is at grade 10 (equivalent with present-day grade 12).'

'So, please take him to my office this afternoon. Invite his father as well.'

That afternoon, Lan Anh took Sơn and his father to my office.

I asked Ms. Trần Thúy Huệ, the head of the natural sciences group, to measure Sơn's knowledge level while I was having the following conversation with his father.

'You usually tutor Son in mathematics at home, don't you?' I asked.

'Oh, no. I have just come back from Bulgaria on business,' Sơn's father said. 'My wife is a pharmacist. She is always busy and cannot remember pre-intermediate mathematics knowledge to tutor our son. Sometimes, my daughter...'

'She is definitely good at mathematics, isn't she?'

'No. In fact, my daughter, Bích Thủy, has aptitude for literature.'

'Bích Thủy and Thanh Sơn, how meaningful their names are! So, I'm wondering how your son learned mathematics.'

'He learned by himself, but I don't know exactly when. Maybe, it was more than one year ago when I went on business in Bulgaria. I was told that he made friends with some neighborhood boys at secondary school and borrowed their mathematics books to try doing (the exercises). Also, he read mathematics books for high school students in secret. I say *in secret* as he read such books only when he was alone. Actually, my wife and I do not want him to study too much. We are afraid of its bad effects on his physical health. However, recently, when my daughter and her classmates reviewed lessons for the entrance examination on university. Sơn overheard their exercises and tried to do them in our brick back yard with chalk. I'm sure that his knowledge is not firm. You know, he just learns by imitation. Our son's mathematics aptitude is mighty good, but we, indeed, do not want to prematurely cram unnecessary things into his young brain. The most important thing to us is his health, that he can develop physically like his peers.'

At the table on the right corner of another room, Sơn was doing some testing exercises after fluently answering oral questions from Ms. Huệ.

Being aware of Sơn's story, other teachers gathered in this room and paid attention to the handwriting of Sơn, who was said to have a special aptitude by Ms. Lan Anh.

Sơn wrote slowly (certainly because he is just a second-grade student), but his solutions for the exercises were accurate and complete.

Some minutes later, Ms. Huệ said to me, 'Sơn's mathematical level is equivalent with that of students at the end of high school, especially in algebra. As for geometry, he is not so good as he hasn't learned about geometry's solid theorem.'

Holding Sơn in my arms, I said to his father, 'Congratulations! You're a happy father. Tomorrow, I will report Sơn's special ability to the Education Department.'

Having permission from the Education Department and Sơn's parents, we suggested that Sơn attend Ms. Huệ's Mathematics class 7 (equivalent with present-day class 9). Ms. Huệ said that Sơn could do the algebra exercises very fast, even faster than senior students in the class. His solutions were generally concise, sometimes unfamiliar, and different from the samples in books. When being asked to perform his solution on the class board, Sơn showed his embarrassment due to the fact that he had not gotten used to writing on the big board. However, his solutions were always well done. It could be said that, in algebra, Sơn was the best student in the class. As for geometry, Sơn wasn't able to keep up with some senior students yet. He didn't have the firm basic knowledge, but after some lessons and listening to Ms. Huệ's lectures, he demonstrated a rapid understanding.

At the end of this school-year, we found that such a large amount of mathematical knowledge did not overload Sơn, so we decided to ask the Education Department's to permit Sơn to skip grade 3 and go directly to grade 4 (equivalent with present-day grade 6). Afterwards, Sơn moved to Tô Hoàng B Mathematics Gifted High School. I retired after that...."

A Fifteen-Year-Old IMO Gold Medal Winner

Sơn passed the entrance exam and became a student at the Mathematics Gifted High School of the Hà Nội University (present-day Việt Nam National University, Hà Nội).

In 1984, he was just fifteen years old. It was the first time he took part in IMO in Prague. He got a gold medal with the absolute score of 42/42 and became the absolute winner.

At that time, Mr. Tạ Quang Bửu, a Vietnamese leading professor, commented, "There are more and more fifteen-year-old students winning high rewards in international mathematical olympiads. How excellent, intelligent and creative they are!"

Professor Tạ Quang Bửu added, "At present, there is no criterion to measure difficult and easy levels of mathematical exercises for students in the last year of high school. However, from our point of view, in this International Mathematical Olympiad the winner did overcome an extremely difficult exercise."

After one-year learning the Russian language in Hà Nội, Đàm Thanh Sơn moved to study at the Physics Department at Moscow State University, the most well-known university in Russia at that time. Since he was very young, Sơn wished to become a theoretical physicist. To make the dream come true, he had to be really good at mathematics. Perhaps, Son was influenced by his uncle, Đàm Trung Đồn, a physics professor.

Having graduated from a university with a high distinction degree, Đàm Thanh Sơn prepared for his doctorate thesis. Professor Valery Rubakov, who is a world-famous physicist and the chief scientist at Moscow Nuclear Research Institute, was Sơn's supervisor. However, the collapse of the Soviet Union put government employees, including scientists, into material difficulties. Realizing that they were great difficulties for the promising talented physicist, Professor Trần Thanh Vân from Paris sent him a letter to encourage Đàm Thanh Sơn to continue pursuing scientific research. He expressed his willingness to give Đàm Thanh Sơn financial aid. Sơn's parents expressed their great gratitude towards Professor Trần Thanh Vân for his kindness but refused to receive his financial aid. They confirmed that they were able to support Sơn's studies by themselves.

On the first *Rencontres du Vietnam* (Việt Nam Meetings) on particle physics and astrophysics in Hà Nội at the end of 1993, Professor Trần Thanh Vân gave Doctor Đàm Thanh Sơn, twenty-four years old at that time, financial assistance to come back to Việt Nam from Moscow to attend and present his scientific report. As I recall, Đàm Thanh Sơn attended the following meetings of *Rencontres du Vietnam* held in 1995, 1999, 2000, and 2004. He was also invited to participate in the Scientific Program Committee together with other world-known scientists many times. Thanks to Professor Trần Thanh Vân's introduction, I had the chance to talk with Professor Valery Rubakov about Professor Đàm Thanh Sơn. This Russian physicist had great compliments for Sơn. He said that although he really wanted to keep Đàm Thanh Sơn working in Moscow, it was him who encouraged Sơn to go to America for his bright development.

In earlier years in America, Đàm Thanh Sơn worked in Professor Lý Chính Đạo's research team at the University of Columbia, New York. Mr. Lý Chính Đạo, whose English name is Tsung-Dao Lee, is a very well-known Chinese American scholar. In 1957, he shared the Nobel Prize with another Chinese American professor, Chen Ning Yang (Dương Chấn Ninh), for their work on parity non-conservation of weak interaction, which caused a widespread phenomenon in the physics field at that time. Now he is about eighty years old. He really likes this young talented Vietnamese physicist and wants him to stay and work at the university for long term. However, in the end, Sơn said good bye to him and moved to work at the University of Washington in Seattle on the bank of the Pacific Ocean. It is warmer there and more suitable for a man coming from a tropical region.

When talking with some journalists in the second *Rencontres du Vietnam* held in Hồ Chí Minh City in 1995, Professor Trần Thanh Vân said, "As we all know, nowadays in American and European countries, Chinese and Japanese people in the fields of natural sciences are no longer looked down on. Hopefully, one day in the future, Vietnamese people can attain such a position. Together with Professor Nguyễn Văn Hiệu in the organization of *Rencontres du Vietnam*. I wish to create an opportunity for young Vietnamese physicists to meet and acquire knowledge from world leading scholars. Hopefully, this will help them keep up with their research. And hopefully, among them, there will appear outstanding talented physicists. At that time, Vietnamese people will be able to demonstrate and affirm their scientific intelligence to the world."

The IPhO (International Physics Olympiad) which was held in Hànội in July 2008 attracted participants from eighty-two countries and territories. Deputy Prime Minister and Minister of Education and Training Nguyễn Thiện Nhân was the leader of the National Steering Committee that is in charge of organizing examinations.

Three Vietnamese-original physicists, including Professor Trần Thanh Vân (Université Paris-Sud), Professor Trương Nguyễn Trân (École Polytechnique de Paris), and Professor Đàm Thanh Sơn (University of Washington in Seattle), were invited to join the organization committee. Among them, Professor Đàm Thanh Sơn was the youngest.

NGÔ BẢO CHÂU – A GLORIOUS PEAK OF MATHEMATICS

He was born in 1972 in Hà Nội and studied during the trying post-war years. However, Ngô Bảo Châu successively won two gold medals in the International Mathematical Olympiad. He became a professor at Paris-Sud University (Université Paris-Sud) and was recognized as a concurrent professor at the Hà Nội Institute of Mathematics at the age of thirty-three. He is presently the youngest professor in Việt Nam.

He successively achieved many international mathematical awards such as the Clay Research Award in America (2004), the Oberwolfach Prize in Germany (2007), and the Sophie Germain Prize from the French Academy of Sciences (2008).

The Princeton Institute for Advanced Study (IAS; Princeton) invited him to go to America to work long term.

In spite of eighteen years of living far away from the motherland, he still carries an ordinary passport issued by the Socialist Republic of Việt Nam.

Born in the Hails of Bombs

Ngô Bảo Châu was born in 1972 when the American president Richard Nixon ordered strategic aircrafts B-52 to hail bombs down on Hà Nội and his father was about to head for the firing line of Quảng Trị in the central part of Việt Nam. He was an only child and grew up in an intellectual family. His father is Ngô Huy Cẩn, a professor and a scientific doctor of mechanics who worked at the Institute of Mechanics (belonging to the Việt Nam Academy of Science and Technology). His mother is Trần Lưu Vân Hiền, an associate professor and a medical doctor who worked at the National Hospital of Traditional Medicine. It can be said that the "hereditary trade" of his family was science.

The wealth of his family does not lie in banknotes or gold, but in books and a desire for knowledge. This has been handed down from generation to generation. His ancestors belonged to the family of "literature." Among them, some were appointed to be the directors of academies that were specialized in teaching Confucian books to the princes of the Nguyễn dynasty.

Ngô Bảo Châu attended the mathematics specialized classes of Hà Nội in Trưng Vương School. Mr. Tôn Thân, one of his mathematics teachers, was well-known as an excellent teacher in Hà Nội. Mrs. Trịnh Bích Ba, his literature teacher, left deep impressions during his teenage years.

After leaving Trưng Vương School, he passed the entrance exam into the Mathematics Gifted Senior High School of Hà Nội University (present-day Mathematics and Informatics Gifted Senior High School, University of Natural Sciences, Việt Nam National University, Hà Nội). A great number of famous Vietnamese mathematics teachers worked at this school for decades. They included Phan Đức Chính, Nguyễn Văn Mậu, Lê Đình Thịnh, Phạm Văn Điều, and Phạm Tấn Dương.

Laozi, a brilliant thinker of the ancient East, wrote, "An a-thousand-mile journey is created by the very first step." Ngô Bảo Châu has set forth firm and correct initial steps. Thanks to training in high-quality specialized classes, in the summer of 1988, at the age of sixteen, and in the eleventh grade, he was selected into the national team of mathematically gifted students to take part in the International Mathematical Olympiad in Canberra, the capital city of Australia. Although he was one of the youngest participants, he reached the top. He won the gold medal with a maximum mark of 42/42. Obviously, he showed an excellent ability in mathematics.

In the following summer of 1989, he was one more time chosen to be a member of the national team of mathematically gifted students for the IMO that took place in Brunswick. Once again, he won the gold medal.

Reaching the Peak in Magnificent Paris

After finishing high school, he took a class in the Hungarian language to prepare for tertiary education in Budapest. Unexpectedly, the Velvet Revolution occurred in Eastern Europe. The new government stopped granting scholarships to Vietnamese students.

Through the introduction of Professor Nguyễn Văn Đạo, Professor Paul Germain, the General Secretary of the French Academy of Sciences at that time, recommended to the French Government that Ngô Bảo Châu should be exceptionally granted a scholarship for Université Paris 6.

For either an excellent French or Vietnamese student, being admitted to the this university was a great satisfaction. Yet, it was not enough for Ngô Bảo Châu. He always tried to reach the highest goal. Two years later, he took an entrance exam into the post-graduate program of École Normale Supérieure – the most well-known university in France where many famous French scholars and some excellent Vietnamese people of the previous generations such as Hoàng Xuân Hãn, Lê Văn Thiêm, Trần Đức Thảo, and Trần Thanh Vân had been trained.

He successfully defended his Ph.D. thesis at the age of twenty-five and then his thesis of habilitation (equivalent to Doctor of Science) at the age of thirty.

Although dedicated to mathematics, he is not just a robot calculating day and night. On the contrary, as observed by his friends, he is a person with a sophisticated soul, thoughtfulness, and a sense of beauty with a love for ballets such as *Polonaise* and *Mazurka* by Chopin, Hàn Mạc Tử's unworldly poems, and Quang Dũng's beautiful poetry. He is also absorbed in reading brand new novels like *The Abandoned Capital* and *The Citadel of Ghosts* written by Jia Pingao, a modern Chinese writer.

Every afternoon, the young father often caressingly takes the hands of his little daughters and strolls leisurely along gentle slopes full of colorful wild flowers in Palaiseau, Gif-sur-Yvette, on the outskirts of Paris. He walks and explains the names of those flowers in both Vietnamese and French to them.

In early 2004, at the age of thirty-one, he was asked to take the position of a professor by two big universities in Paris (Université Paris 6 and Université Paris 11). He accepted the proposal of Université Paris 11 out of a desire to live near his three colleagues, Gérard Laumon, Laurent

Lafforgue, and Ngô Đắc Tuấn. Gérard Laumon had instructed Laurent Lafforgue and Ngô Bảo Châu in writing their Ph.D. theses. Both Gérard Laumon and Laurent Lafforgue have already visited Việt Nam and lived in the house owned by Ngô Bảo Châu's parents in Mai Động, Hà Nội, for many days.

Similar to Ngô Bảo Châu, Ngô Đắc Toán was also originally a student in the mathematics specialized class at Hà Nội University and won the gold medal twice in the International Mathematical Olympiad in Canada (1995) and India (1996). He then studied in École Polytechnique de Paris and came in first in the graduation examination. It was Ngô Bảo Châu who introduced him to Laurent Lafforgue, who later helped him with his Ph.D. thesis. (Laurent Lafforgue is presently an academician of the French Academy of Sciences. He is also Ngô Bảo Châu's close friend and a Fields Medal-awarded mathematician.)

A Mathematical Work of Immediate Resounding Success

In April 2004, Ngô Bảo Châu and Gérard Laumon announced a 100-page mathematical work in French entitled *Le lemme fundamental pour les groupes unitaires* (The Fundamental Lemma for Unitarian Groups) in preprint form and posted it onto the internet. This work looked into a topical mathematical issue. It solved a big problem that mathematicians had previously tried to solve for twenty years. It immediately gained a widespread resounding success.

Ngô Bảo Châu was invited to Japan to present his new findings and then to Canada to attend the International

Conference on Automorphic Forms and the Trace Formula at the Fields Institute. Many famous mathematicians from major universities in the world attended the conference. Ngô Bảo Châu was invited to read a report in the first plenary session. After listening to his presentation, Robert Langlands, a mathematician who had introduced the Langlands Program that had attracted the interest of many world famous mathematicians over the past decades, met him right away and invited him to work at Institute for Advanced Study/ IAS in Princeton – a top institute in America where Albert Einstein used to give lectures – for long term.

Removing an Obstacle in the Flow of Mathematics

Thanks to his experience in successfully studying the fundamental lemma of Jacquet, Ngô Bảo Châu bravely started studying that of Langlands. After two years, he achieved a breakthrough in the summer of 2003 when returning to Hà Nội where "red flamboyant flowers shaded the dykes and cicadas sang merrily on the dracontomelum trees." He was returning to visit his parents at a lovely newly-built house in view of the deep blue lake of Thủ Lệ. In the following months, together with some previous findings from Gérard Laumon, he and Gérald Laumon completed proving the fundamental lemma for unitarian groups.

The work of Ngô Bảo Châu and Gérard Laumon successfully proved this lemma, removing a consistent obstacle in the flow of modern mathematics and immediately attracting the attention of international mathematicians. These two authors helped mathematicians overcome an obstacle and go further along the path that Andrew Wiles had gone through when he proved the Taniyama-Shimura

Hypothesis. With the findings of Ngô Bảo Châu and Gérard Laumon, the international mathematician circle has taken one more step toward proving other hypotheses in the Langlands Program and realizing the cherished dreams of many generations of researchers who aimed to find a splendid unity in mathematics.

It was not by chance that Andrew Wiles, "the most well-known mathematician of the twentieth century," came forward and recommended Ngô Bảo Châu and Gérard Laumon as receivers of the Research Award of the Clay Mathematics Institute for the most excellent mathematical work in the world in 2004. And it was not strange that the Americans invited Ngô Bảo Châu to work in their country with a salary of more than 200,000 US dollars per year.

Clay Research Award: An International Honor

In mid October 2004, Ngô Bảo Châu attended the International Conference on Automorphic Forms and the Trace Formula at the Fields Institute in Canada. Together with many world-famous scholars, he was invited to read a report in the plenary session. He demonstrated the work that had been completed and announced on the internet by him and Professor Gérard Laumon. This 100-page work was about the fundamental lemma for unitirian groups, solving a roadblock on the way of developing the automorphic forms theory and gradually realizing the Langlands Program.

Before the conference in Canada, he received an email from James Carlson, President of the Clay Mathematics Institute. The president wrote,

"Dear Professor Ngô,

I'm delighted to inform you that the Clay Mathematics Institute has selected you and Gérard Laumon as recipients of the Clay Research Award to be presented at its annual meeting in Cambridge, Massachusetts, on 5 November. The purpose of this award is to recognize your joint work on fundamental lemma. Last year's recipients were Richard Hamilton and Terry Tao. In the preceding year, the award went to Manindra Agrawal and Oded Schramm.

Our scientific advisory board, James Carlson, Simon Donaldson, Gregory Margulis, Richard Melrose, Yum-Tong Siu, and Andrew Wiles, offers its congratulations.

I would like to discuss with you, at your convenience, what kind of research support the Clay Mathematics Institute might offer you and Laumon in the coming year. The range of possibilities is quite wide, and could include travel, travel for persons with whom you would like to collaborate, organization of workshop, etc.

In addition, I would like to invite you to our meeting in Cambridge on Friday, 5 November to receive award. CMI will pay all travel, lodging, and per diem expenses. If you would like to extend your visit beyond the time of the meeting, before or after, so as to speak with mathematicians here and elsewhere in America, we would be delighted to reimburse you for those expenses as well.

I am also writing to Laumon to inform him of the award, to invite him to attend the meeting, and to receive the award with you. I have invited him to speak about the work that you and he have done on the fundamental lemma.

I look forward to speaking with you soon. Perhaps we can arrange a convenient time for me to call?

Most sincerely,

Jim Carlson

President, Clay Mathematics Institute."

After the conference at the Fields Institute finished, Ngô Bảo Châu returned to Plaiseau with its green pine hills and vast buckwheat fields in the suburbs of Paris where he lives with his wife, Mrs. Nguyễn Bảo Thanh – his classmate from the mathematics specialized class at Trưng Vương School, and his three little daughters. His passport was going to expire. He had to go to the Vietnamese Embassy in Paris to renew it. It was not difficult. Yet, he wondered if he would have enough time to make an entry visa into the United States. In recent years, due to the apprehension of terrorists, granting visas to citizens of Asia and Africa into the United States may last a whole month. He has studied in France since 1990. Despite having lived and worked in Paris for almost fifteen years, he still retains the Vietnamese nationality and carries an ordinary green passport. It is because he obeys to the moral philosophy "being grateful to one's benefactors." He intended to ask Gérard Laumon to stand for both of them and go to Cambridge to receive the Clay Research Award because Gérard Laumon had a French passport and did not need a visa to enter the United States. Almost saying goodbye to all hopes, he was suddenly informed that the Clay Mathematics Institute had asked Senator Edward Kennedy to intervene and call the American Embassy in Paris from Boston. And he got a visa a few hours later.

The Granting Ceremony of the Clay Research Award in 2004 took place simply but solemnly during an annual meeting of the Clay Mathematics Institute at Harvard University (America) on 5 November 2004. By then, only twelve mathematicians in the world had received this award.

In 2005, the Vietnamese government exceptionally recognized Ngô Bảo Châu as Professor and Doctor of Science. At the age of thirty-three, he became the youngest professor in Việt Nam.

After receiving the Clay Research Award, Professor Gérald Laumon was appointed Member of the French Academy of Sciences.

Oberwolfach Prize and Sophie Germain Prize

After receiving the Clay Award in America, Ngô Bảo Châu was awarded the Oberwolfach Prize in the Federal Republic of Germany. This is a mathematical prize that is awarded every three years to one or two mathematicians under thirty-six years old who have outstanding research works in Europe. The Oberwolfach Prize in 2007 went to only Ngô Bảo Châu for his outstanding 188-page work on algebra and number theory. This prize is awarded by the Oberwolfach Foundation and the Oberwolfach Mathematical Research Institute in Germany. The new work by the Vietnamese mathematician working at Orsay University (Université Paris-Sud) aimed to completely solve the fundamental lemma according to the conjecture of Langlands and Shelstad. With exact demonstrations, Ngô Bảo Châu was acknowledged as the leading expert in the joint between algebraic geometry and automorphic forms theory. In 2004, together with Gérald Laumon, he just solved the fundamental lemma for unitarian groups. Presently, he gave an answer to a more general case.

Reading the laudatory speech in the Awards Ceremony held in the beginning of 2008, Professor Rapoport regarded his latest work as "a brilliant achievement."

After that, he also received the Sophie Germain Prize, a prize from the French Academy of Sciences.

Loved and Admired by "Strict" Colleagues

On assessing the success of Ngô Bảo Châu, Professor Ngô Việt Trung, Director of the Việt Nam Institute of Mathematics and Member of the Third World Academy of Sciences, said in an interview, "In the beginning of 2004, Ngô Bảo Châu and Professor Gérard Laumon created a 'great bomb' when announcing breakthrough findings about the fundamental lemma in Langlands Program and made a resound in the international mathematical circle. With these findings, he and Professor Gérard Laumon were awarded the prestigious Clay Award. Then he was invited to be a professor at Paris-Sud University. In 2007, he once again caused a stir when completely solving the fundamental lemma. With such a success, he was invited to work for a long time by the Princeton Institute for Advanced Study where a great number of leading scientists in the world are working."

Then he drew a conclusion, "We have a ground to hope that Ngô Bảo Châu is awarded one of the most prestigious prizes in mathematics – the Fields Medal."

Professor Nguyễn Tiến Dũng from Toulouse University (Université de Toulouse), France, remarked, "If we are proud of Vietnamese bonze drums, we are also proud of a Vietnamese person who has obtained a world-famous mathematical achievement. This person is Ngô Bảo Châu."

From 19 to 27 August 2010, Professor Ngô Bảo Châu attended the International Congress of Mathematicians in Hyderabad, India. In the opening session, he was awarded the Fields Medal by Mrs. Pratibha Devisingh Pratil, President of Indian Republic, and Professor Lászlo Lovász, President of the International Mathematical Union.

NHÀ XUẤT BẢN THẾ GIỚI
Trụ sở chính: 46 Trần Hưng Đạo, Hà Nội, Việt Nam
Tel.: 84.4.38253841 / 38262996 - Fax: 84.4.38269578
Chi nhánh: 7 Nguyễn Thị Minh Khai, Quận 1,
TP. Hồ Chí Minh, Việt Nam
Tel.: 84.8.38220102
Email: thegioi@hn.vnn.vn
Website: www.thegioipublishers.com.vn

TRÍ THỨC VIỆT NAM: NHỮNG GƯƠNG MẶT TIÊU BIỂU

Chịu trách nhiệm xuất bản:
Trần Đoàn Lâm

Biên tập:
Quách Ngọc Anh

Hiệu đính:
Nathan Morrow

Trình bày:
Phạm Trần Hiếu

Bìa:
Trung Dũng

In 800 bản, khổ 14 x 20,5 cm, tại TT Chế bản và In NXB Thế Giới. Giấy chấp nhận đăng ký kế hoạch xuất bản số: 473-2011/CXB/3-66/ThG cấp ngày 18/05/2011. In xong và nộp lưu chiểu năm 2011.